Key Issues in
Environmental Change

Series Editors:

Co-ordinating Editor

John A. Matthews
Department of Geography, University of Wales Swansea, UK

Editors

Raymond S. Bradley
Department of Geosciences, University of Massachusetts, Amherst, USA

Neil Roberts
Department of Geography, University of Plymouth, UK

Martin A. J. Williams
Mawson Graduate Centre for Environmental Studies, University of Adelaide, Australia

Preface to the series

The study of environmental change is a major growth area of interdisciplinary science. Indeed, the intensity of current scientific activity in the field of environmental change may be viewed as the emergence of a new area of 'big science' alongside such recognized fields as nuclear physics, astronomy and biotechnology. The science of environmental change is fundamental science on a grand scale: rather different from nuclear physics but nevertheless no less important as a field of knowledge, and probably of more significance in terms of the continuing success of human societies in their occupation of the Earth's surface.

The need to establish the pattern and causes of recent climatic changes, to which human activities have contributed, is the main force behind the increasing scientific interest in environmental change. Only during the past few decades have the scale, intensity and permanence of human impacts on the environment been recognized and begun to be understood. A mere 5000 years ago, in the mid-Holocene, non-local human impacts were more or less negligible even on vegetation and soils. Today, however, pollutants have been detected in the Earth's most remote regions, and environmental processes, including those of the atmosphere and oceans, are being affected at a global scale.

Natural environmental change has, however, occurred throughout Earth's history. Large-scale natural events as abrupt as those associated with human environmental impacts are known to have occurred in the past. The future course of natural environmental change may in some cases exacerbate human-induced change; in other cases, such changes may neutralize the human effects. It is essential, therefore, to view current and future environmental changes, like global warming, in the context of the broader perspective of the past. This linking theme provides the distinctive focus of the series and is mentioned explicitly in many of the titles listed opposite.

It is intended that each book in the series will be an authoritative, scholarly and accessible synthesis that will become known for advancing the conceptual framework of studies in environmental change. In particular we hope that each book will inform advanced undergraduates and be an inspiration to young research workers. To this end, all the invited authors are experts in their respective fields and are active at the research frontier. They are, moreover, broadly representative of the interdisciplinary and international nature of environmental change research today. Thus, the series as a whole aims to cover all the themes normally considered as key issues in environmental change even though individual books may take a particular viewpoint or approach.

John A. Matthews (Co-ordinating Editor)

Titles in the series

Atmospheric Pollution: an Environmental Change Perspective (Sarah Metcalfe, Edinburgh University, Scotland)

Biodiversity: an Environmental Change Perspective (Peter Gell, Adelaide University, Australia)

Climatic Change: a Palaeoenvironmental Perspective (Cary Mock, University of South Carolina, USA)

Environmental Change at High Latitudes: a Palaeoecological Perspective (Atte Korhola and Reinhard Pienitz, Helsinki University, Finland & Laval University, Québec, Canada)

Environmental Change in Drylands (David Thomas, Sheffield University, UK)

Environmental Change in Mountains and Uplands (Martin Beniston, University of Fribourg, Switzerland)

Glaciers and Environmental Change (Atle Nesje and Svein Olaf Dahl, Bergen University, Norway)

Natural Hazards and Environmental Change (W.J. McGuire, C.R.J. Kilburn and M.A. Saunders, University College London, UK)

Pollution of Lakes and Rivers: a Palaeoecological Perspective (John Smol, Queen's University, Canada)

Wetlands and Environmental Change (Paul Glaser, Minnesota University, USA)

Cultural Landscapes and Environmental Change

Lesley Head

School of Geosciences, University of Wollongong, Australia

A member of the Hodder Headline Group
London
Co-published in the United States of America by
Oxford University Press Inc., New York

First published in Great Britain in 2000 by
Arnold, a member of the Hodder Headline Group,
338 Euston Road, London NW1 3BH

http://www.arnoldpublishers.com

Co-published in the United States of America by
Oxford University Press Inc.,
198 Madison Avenue, New York, NY10016

The advice and information in this book are believed to be true and
accurate at the date of going to press, neither the author[s] nor the publisher
can accept any legal responsibility or liability for any errors or omissions.

British Library Cataloguing in Publication Data
A catalogue record for this book is available from the British Library

Library of Congress Cataloging-in-Publication Data
A catalog record for this book is available from the Library of Congress

ISBN 0 340 73113 3 (hb)
ISBN 0 340 73114 1 (pb)

1 2 3 4 5 6 7 8 9 10

Production Editor: Wendy Rooke
Production Controller: Bryan Eccleshall
Cover Design: Mouse Mat Design

Typeset in 10pt Palatino by Saxon Graphics Ltd, Derby
Printed and bound in Malta by Gutenberg Press

What do you think about this book? Or any other Arnold title?
Please send your comments to feedback.arnold@hodder.co.uk

To the memory of Toni O'Neill
(1945–1999)
environmental scientist of distinction,
teacher and friend to many,
lover of landscapes and of cultures

Contents

Figures

Illustrations

Tables

Boxes

Preface:
quarrying yams – perspectives from the edge

When Aboriginal women in the Keep River area of north-western Australia go for yams, they target well-known, named patches, mostly small rocky outcrops vegetated with dry monsoon rainforest. Tubers of the two main yam species, *Dioscorea transversa* and *D. bulbifera*, may be over a metre below the ground surface, and several hours may be invested in the pursuit of a single one (Illustration 0.1a). This work also involves movement of large quantities of rock and soil, leaving holes over a cubic metre in size. The holes are a clear sign of human activity; recently worked ones lack sedimentary or leaf litter infilling. Older holes, partly filled in with slope wash (Illustration 0.1b), are recognized by contemporary custodian Biddy Simon as places where people have dug in the past: 'you look all them hollow, all them hole, that where people bin sittin down, down old time … they bin sitting down and coming in from all around' (J. Atchison, in prep.). Atchison has argued that the holes may serve an inverse function to garden mounds. Instead of hilling up the soil profile above the ground surface, the holes create a deeper stone-free soil structure for replanted yams to grow in.

At one site, Milyoonga, the foot of the rocky yam slope is a stone quarry. Over an unknown time period people have visited the site to select source materials for stone artefacts. Cores from which flakes have been removed, as well as some of the flakes themselves, remain. Soil samples from the yam holes contain both flakes, and stone pieces shattered by digging. While this evidence of human activity is not as dramatic as the rock art in the surrounding hills, and would not be recognized by as many people, it is a landscape transformed by people.

For many years researchers have debated whether surviving rainforest patches in northern Australia are relics of past, more mesic climates, and whether Aboriginal burning over millennia has contributed to their patchiness. The dry rainforest areas providing yam habitat in the Keep River are patchy in distribution because so is the present location of rocky outcrop in the surrounding sandy and estuarine plains. At times of lower sea level (prior to 6000 years ago) or less extreme dry seasons (8000–4000 years ago) the configuration may have been quite different. Within the dry rainforest vegetation, there are two floristic elements which probably responded differently; the opportunists 'with a capacity to exploit new ecological opportunities', and the refugia taxa whose existence is more precarious (Russell-Smith and Dunlop, 1987: 230). Over many thousands of years Aboriginal yam gardens may have differentially affected the opportunities available to these two elements.

In the Keep River region, use of yam patches is part of a broad pattern of regular visitation to and connection between sites, showing a network of linkages across the landscape. There have been both continuities and changes in these linkages over the past hundred years of incursion by Euro-Australian pastoralism. The distribution of certain stone artefacts away from geologically

ILLUSTRATON 0.1 (a) Biddy Simon digs for yams in the Keep River area, Northern Territory, Australia. (b) Yam-digging holes in dry rainforest in the Keep River area, Northern Territory, Australia (by permission of Jennifer Atchison)

identifiable sources can tell us about the operation of exchange networks in the past, some of which continue. Miriuwung and Gajerrong owners define their ownership rights over this land in terms of their relationships to one another, to now deceased kin and to ancestral Dreaming figures. These rights are expressed in stories which link yam hills, porcupine places, stone quarries and ochre sources across the landscape. The retelling of the stories in suitably edited form to visitors affirms the authority of the teller to speak for country.

Following the dislocations of European pastoralism in the past century, custodianship rights and responsibilities have also been given to Murinpatha people from further east. Favoured sites for fishing, turtle and goanna collecting are visited on a regular basis, integrated now with trips to town for market goods. High-quality ochre is collected from particular quarries, as it has been for millennia, and is now used in paintings for sale to tourists.

The different readings of the Keep River and surrounding landscape and its past come into collision in discussions of the area's social and environmental future; this is a contested landscape. The governments of Western Australia and the Northern Territory want to expand irrigated agriculture from the neighbouring Ord basin onto the black soil plains of the Keep River (Illustration 0.2). The sandstone outcrops – yam, ochre, stone and rock art places – would protrude above plains of sugar and cotton. Miriuwung-Gajerrong people have successfully claimed native title to a large area of unalienated Crown land in the region, making them key stakeholders in decisions

ILLUSTRATION 0.2 Irrigated area near Cave Spring Range, north of Kununurra, Western Australia. This is near the northernmost extent of the current Ord Stage I scheme, due for expansion in the next few years

about environmental futures. The decision is currently under appeal. Meanwhile, tourists flying into Kununurra airport are welcomed to 'Australia's last frontier', a conceptualization that jars with both occupied Aboriginal space and intensive agriculture.

Culture and nature, symbolism and materiality, past and present are clearly intertwined in this example, as they are at different scales in all environmental issues. In starting with an example that most readers would find exotic and faraway, I risk diverting attention from more everyday, ordinary cultural landscapes. (Although to the Miriuwung-Gajerrong, of course, it is the cities and suburbs of eastern Australia or the North Atlantic that are faraway, foreign and wild.) In the coastal suburb in which I live, the line between culture and nature is also blurred in subtle and interesting ways. It is often noted that most white Australians cling to the edges of the continent. It is the environment in which we consider ourselves to feel most at home. Swimming out behind the breakers, I can imagine myself in a totally natural landscape. The offshore wind drives up spray from the cresting waves and blurs the beach. From here, the deep green wave backs curve into the greyer forest greens of the escarpment behind, absorbing the houses and hiding the sand and foam of the edge. My illusion is obviously false, and can be maintained only by squinting. Apart from anything else, I know that the creeks draining the escarpment bring motor oil, dog faeces and plastic bags to the sea.

In the early decades of the twentieth century, people cut swimming pools into the rocky headlands along this coast (cover illustration). When the tide is just so, a swimmer can imagine herself on the edge of the horizon. On the strandline after a storm, the plastics can take on the weathered, at-home look of driftwood. Is that a piece of broken surfboard, or a cuttlefish? the perished corner of a rubber flipper, or a desiccated sponge? The betweenness of the coast, its constant movement and the multiple layers of its edges, provide useful images for the thinking in this book. In our daily lives the interpenetration and mutual constitution of people and their environment is a lived experience. Why do we continue to feel the need to separate them analytically?

The coast is also, of course, an important locus of environmental change. Swimming behind the breakers, I can have another false illusion, that I am on the ocean. Though deep in human terms, the waters beneath me are geologically speaking not oceanic at all; the edge of the continental shelf is somewhere out near the horizon. Ten thousand years ago and beyond, people crossed a more extensive coastal plain, above which I now float, to a more distant ocean. In the event of a future sea level rise or increased storminess, the expensive real estate on the headlands will fall into the sea. A tsunami triggered by volcanic activity or a submarine landslide out in the Pacific would have a more sudden and dramatic impact. Likely future changes will be the product of both human activity and processes over which we have no control. And on the other side of the continent, where the shelf is broad rather than narrow, landscape changes associated with sea-level rise and fall are even more dramatic (Illustration 0.3).

ILLUSTRATION 0.3 Coastal and estuarine landscapes from five rivers lookout, near Wyndham, Western Australia. At times of lowest sea level, such as at the height of the last ice age 18,000 years ago, the coastline would have been hundreds of kilometres away to the north-west

There are many edges in this book. The lines between culture and nature, science and the humanities, people and their environment, north and south, are as problematic and shifting as the continental edge from which I write. Our challenge is not to ink in the lines but to use the tensions creatively in order to see more clearly where we stand in space and time.

Acknowledgements

As will be apparent from the references, many people have stretched my reading and thinking on the issues discussed in this book. For stimulating discussion of specific matters, access to unpublished material, comments and advice (not always taken) on this and related manuscripts, and/or timely encouragement, I particularly thank Michael Adams, Jenny Atchison, Dave Bowman, Tony Brown, Richard Fullagar, Peter Gell, Simon Haberle, Peter Kershaw, Ruth Lane, Matt McGlone, Les Rowntree, Biddy Simon, Paul Taçon, David Taylor, Robin Torrence, David Trigger and Gordon Waitt.

At the University of Wollongong, important research support over a number of years has been provided by the Quaternary Environments Research Centre, and particularly by colleagues Ted Bryant, Allan Chivas, Colin Murray-Wallace, Gerald Nanson, David Price and Colin Woodroffe. A grant from the university's Environment Research Institute enabled Ruth Lane to assist with library research on this project. In this regard I thank also the university's inter-library loans staff. The Faculty of Science and the university provided the period of study leave which made the writing of the book possible.

The Australian Research Council and the Australian Institute of Aboriginal and Torres Strait Islander Studies have been important sources of funding support. For logistic support and inspiration of various kinds I thank Wendy Beck, Clare Boyd-Macrae, Sue Bromham, Judith Field, Carol Head, Robert Irving, Jo McDonald, Penny Podimatopoulos, Sue Pritchard and Margaret Sheil. I thank Luciana O'Flaherty and Wendy Rooke of Arnold for their encouragement and efficiency. My greatest debt is to Richard, Hugh and Hannah Fullagar, who keep me grounded in the present.

The author and publishers would like to thank the following for permission to use copyright material in this book:

Text

World Heritage Centre, UNESCO, Paris, for extracts from the World Heritage Convention and the List of World Heritage Sites; Cambridge University Press for table 1.3 from B.L. Turner *et al.* (eds) (1990) *The Earth as transformed by human action;* Taylor & Francis Ltd, PO Box 25, Abingdon, Oxfordshire OX14 3UE, UK, for table 2 from M. Jones and K. Daugstad (1997) Usages of the 'cultural landscape' concept in Norwegian and Nordic landscape administration. *Landscape Research* **22**(3): 267–81.

Figures

The University of Wisconsin Press for figure 8.1, figure 8.2 and map 7.1 from K.S. Zimmerer and K.R. Young (1999)(eds) *Nature's geography;* Kluwer Academic Publishers for Table from J. Decher (1997) Conservation, small mammals, and the future of sacred groves in West Africa. *Biodiversity and Conservation* 6: 1007–26; Blackwell Science Inc. for figure 2 from R.J.Hobbs and D.A. Norton (1996) Towards a conceptual framework for restoration ecology. *Restoration Ecology* **4**: 93–110, and figure 2 from J. Aronson and E. Le Floc'h (1996) Hierarchies and landscape history: dialoguing with Hobbs and Norton. *Restoration Ecology* **4**: 327–33; David Taylor for 'Summary of variations in sediments and signals of human activity from five locations in western Uganda'; Cambridge University Press for figure 2.2 from J. Fairhead and M. Leach (1996) *Misreading the African landscape*; University of Minnesota Press for figure 11.7 from H.E. Wright, Jr (ed.) (1983) *Late-Quaternary environments of the United States, volume 2, The Holocene*; Peter Kershaw for 'Charcoal peaks in long continuous records from the Australian region in relation to the marine oxygen isotope (SPECMAP) record' and 'Relative importance of burning in time slices from the Holocene'; Elsevier Science for a figure from S.G. Haberle, G.S. Hope and W.A. Van der Kaars (in press) Biomass burning in Indonesia and Papua New Guinea. *Palaeogeography, Palaeoclimatology, Palaeoecology*, and figure 3 from A. Ghaffar and G. Robinson (1997) Restoring the agricultural landscape: the impact of government policies in East Lothian, Scotland. *Geoforum* **28**: 205–17; the American Association for the Advancement of Science for figures 1 and 2 from P.M. Vitousek, H.A. Mooney, J. Lubchenco and J.M. Melillo (1997) Human domination of Earth's ecosystems. *Science* **277**: 494–9; John Barrett for 'Four views of Silbury Hill' from *Fragments from antiquity: an archaeology of social life in Britain, 2900–1200 BC*; Blackwell for figure 8.2 from N. Roberts (1989) *The Holocene: an environmental history*; Australian Archaeology for two figures from S. Haberle (1998) Dating the evidence for agricultural change in the highlands of New Guinea: the last 2000 years; Springer-Verlag for figure 2 from M. Sykes (1997) in B. Huntley, W. Cramer, A.V. Morgan, H.C. Prentice and J.R.M. Allen (eds) *Past and future rapid environmental changes: the spatial and evolutionary responses of terrestrial biota*; Michael Leunig/*The Age* for 'Past and future'; Matt McGlone for 'Past and future vegetation cover of New Zealand'; Taylor & Francis Books Ltd for 'Spatial and temporal scales of environmental change' and 'July temperatures 18,000 years ago, 9,000 years ago

and present', both from R.J. Huggett (1997) *Environmental change: the evolving ecosphere*, Routledge, figs 1.2, 4.4, 'The prehistoric archaeological landscape at Loughcrew, Ireland', from G. Cooney (1999) Social landscapes in Irish prehistory, in P. Ucko and R. Layton (eds) *The archaeology and anthropology of landscape*, Routledge, p. 53, 'Cultural landscape in material form, gendered and divided according to its prestige or everyday purpose', 'Vuollerin 6000 år: not-so-hidden messages about ethnicity' and 'Lule River, upper valley in northern Sweden, showing the boundaries of various national parks which constitute Laponia World Heritage Area', all from I.-M.Mulk and T. Bayliss-Smith (1999) The representation of Sami cultural identity in the cultural landscapes of northern Sweden, in Ucko and Layton, figs 24.2, 24.3, 24.6 and 24.1, 'Schematic map of anthropogenic vegetation changes in West Africa during the twentieth century', from J. Fairhead and M. Leach (1998) *Reframing Deforestation: global analyses and local realities*, map 1.1.

Illustrations

Jennifer Atchison for 'Yam digging holes in dry rainforest, Keep River area, Northern Territory, Australia'; Atholl Anderson for 'Excavation in progress, moa-hunting site, Shag River mouth, New Zealand' and 'Excavated moa butchery area, Hawksburn site, Central Otago, New Zealand'; Gordon Waitt for 'Pagoda, Namsan Mountains park, near Kyongju, South Korea', 'Agricultural landscape with hedgerows, West Lothian, Scotland', 'Ploughed land to edge of hedgerow, East Lothian, Scotland', 'Juxtaposition of old and new, Seoul, South Korea' and 'Buddha relief on stone, Namsan Mountains Park, near Kyongju, South Korea'; Richard Fullagar for 'Tourists watch the sunset at Uluru' and 'Cupules at the Jinmium rock-shelter site'; Gerald Nanson for 'River rehabilitation: pin groynes designed to accumulate sediment on the bar beneath the opposite bank and prevent erosion of the bank, Taylors Arm, Nambucca catchment, Australia'; Mary-Jane Mountain for 'Remains of circular stone monument, Komakino, near Aamori, Japan', 'Sannai Murayama site, Aamori, Japan' and 'Visitors look at excavated ground surface, Sannai Maruyama site, Aamori, Japan'; Matt McGlone for 'Bracken, tussock grassland and forest remnants, Port Hills, Canterbury, South Island, New Zealand'.

Part I

OVERVIEW

1

Contingent constructions: cultural landscapes and environmental change

1.0 Chapter summary

As human influence on the Earth and its processes increases, we face the profound paradox that most of our intellectual weapons in the environmental area have maintained a separation of humans and nature. Can we rethink people into nature in such a way that we can better manage the Earth? Clear distinctions between cultural and natural landscapes have been challenged by recent work in both the sciences and the humanities. There are grounds for optimism in the convergence of independent lines of evidence about human interactions with environment. This convergence will be summarized in the umbrella concepts of *contingency* and *construction*.

Challenges to apparently self-evident truths such as pristine nature have alarmed some environmental scientists, who are worried about relativism and the loss of a basis for action. However, the challenge needs to be engaged with, and in fact provides some creative new directions. Much of our thinking on these issues is constrained by a number of dualisms within Western thinking. It is possible and necessary first to reorient those dualisms, for example by bringing together thinking in palaeoecology and cultural geography. Second, it is important to draw on the knowledge traditions of indigenous and other non-Western peoples to show other possible ways of

thinking about human–nature interactions. This is not to imply, however, that using knowledge of the past as a resource in contemporary environmental management is a simplistic process of either 'going backwards' or 'learning lessons'. Rather, we need to critically analyse not only the way we produce knowledge about the past, but also how that knowledge is used in today's debates. I outline the theoretical underpinnings and operational guidelines which will emerge in the discussions in the book.

1.1 Disappearing nature and the lessons of the past

Physical geographers and others are having trouble with culture – it is spreading in both space and time. Cultural landscapes are usually defined within the environmental sciences with reference to an antithesis, namely natural landscapes. At least implicit in this understanding are ideas about human transformations of the Earth and of landscapes containing relics of past human activity. Most physical geographers writing about cultural landscapes begin from the perspective of the physical environment itself; thus a cultural landscape is seen as one materially modified by people, a natural one is not. However, this distinction between cultural and natural landscapes has been rendered problematic by recent work. Just as earlier views of natural landscapes required revision when the profound effects of agricultural activity became apparent (Faegri, 1988), so researchers now routinely consider – and debate with great spirit – the impacts of hunter-gatherers on biotic and physical landscapes (Dodson, 1992). Hunter-gatherer impacts are now understood as being both earlier and subtler than traditionally thought. The pristine baseline – the 'natural landscape' – is a mirage, receding as it is approached. Even if found, in most parts of the world it is of limited value as a state against which later impacts can be assessed, as it existed under different boundary conditions, particularly of climate.

Human geographers and others are also having trouble with culture – it is a complex, multidimensional idea that has both material and symbolic expression. Cultural geographers, for example, recognize landscapes as 'transformations of social and political ideologies into physical form' (Duncan and Duncan, 1988: 25). In a somewhat parallel trend, but on a different timescale, archaeologists are using the term 'social landscape' to understand long-term change. In this understanding, 'the use of different parts of the landscape is seen to be structured by the needs of the social system' (Gosden, 1989: 47). While archaeologists by definition work with material evidence, not all dimensions of landscape have material expression. Other writers have used the notion of landscape as text, 'a medium to be read for the ideas, practices and contexts constituting the culture which created it' (Ley, 1985: 419). The metaphor of text may be problematic for thinking about the conceptual landscapes of peoples without writing. The important point here is that landscapes are 'read' very differently by different individuals and communities. In turn, this throws into question the ways in which we as researchers analyse and portray other cultures, particularly as many of the 'others' discussed in this book are separated from us by time as well as space. This thread questions the very distinction between the material and the symbolic, seeing 'all artefacts … as texts or representations' (Daniels, 1989: 198).

Trouble with culture means trouble with nature. An important body of literature is that on the cultures of nature, which has provided critical re-examination of the diverse ways in which social groups conceptualize nature. In this literature nature is not so much a mirage as a category constructed within specific social contexts. As those contexts differ, so do nature and the natural.

These two mostly separate debates are both challenging and, sometimes, alarming for scientists working on contemporary environmental management issues. Surely the only disappearing nature that we have time to talk about is that shrinking under the accelerating onslaught of deforestation, land degradation, increasing urbanization and population growth? Satellites that facilitate the instantaneous communication upon which the globalized economy depends are harnessed to track and predict the path of its voracious appetite. We remotely sense problems we cannot even see, such as ozone holes and carbon dioxide concentrations.

Despite the apparent urgency, we sense that the past has something to tell us. We look back, finding evidence of climate change, fire, sea-level rise and fall, forest change and movement, plant and animal extinctions. Our looking back is increasingly dependent on that same high technology; we can measure the minuscule and incorporate it into modelling. Thus we build up scientific stories of long-term environmental change, but how precisely are they to help us in our present and future decisions? Is the past there to provide a wake-up call by showing the trajectory of human impacts on the Earth? It is an epic story, but will it be one of redemption or damnation? How is the past to save us, when we are all too conscious that we cannot go back to spears and loincloths?

There are two great clichés in the literature which seeks to harness creatively the long-term record to solve contemporary environmental problems: that research needs to be interdisciplinary; and that we can manage the future only if we understand the past. But how will it help? We cannot go backwards; we cannot just 'learn the lessons' of the past and easily effect transformation of our own social context. As well as critically examining the long-term record, we need to critically examine how it can be usefully applied.

This book is an attempt to outline the ground on which we can do just that. Before I summarize the theoretical underpinnings and operational guidelines of such an outline, it is helpful to clarify the disciplinary perspective in which I am operating.

1.2 Disciplines and dualisms

Perhaps, then, everything is cultural and nothing is natural? For environmental management and conservation developed within a Western scientific tradition this is a problematic assertion. In a practical sense, the demise of nature encapsulates for us all our contemporary environmental problems. At a deeper level, however, the assertion challenges the dualisms deeply embedded in Western thinking – culture/nature, mind/body – and the intellectual tools with which we think about them – humanities/sciences. Within the discipline of geography, the dualism is expressed in the division between human and physical.

Most attempts at environmental interdisciplinarity focus at a zone we tend to think of as the human–physical interface of a continuum. For example, within geography we tend to conceptualize the subfields of the discipline in terms of a continuum between the most physical and the most human (Figure 1.1a). (I do not wish to suggest that this subject matter is only the preserve of geography–far from it. However, the dilemmas within the disciplinary tradition of geography illustrate these broader issues well.) A number of other features of this debate are also conceptualized along a continuum. For example, the positivist/empiricist tradition of transformed cultural landscapes lies at one end, and the social constructivist tradition of landscape as representation at the other (Figure 1.1b). The temporal dimension is often thought of in the same way, with deep time being dominated by nature, and human agency gradually increasing until the present (Figure 1.1c). In some ways this is also a continuum between the 'sciences' on the left and the 'humanities' on the right, although that summary submerges to too great an extent the historical-descriptive tradition of much environmental change research. There is also a tendency for

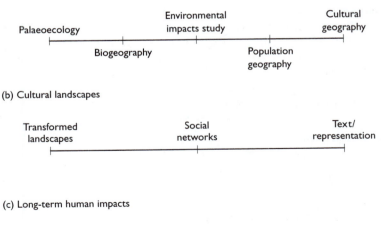

(a) Geography

Palaeoecology   Environmental impacts study   Cultural geography

Biogeography   Population geography

(b) Cultural landscapes

Transformed landscapes   Social networks   Text/ representation

(c) Long-term human impacts

Agency of nature   Accelerating enculturation and damage   Human domination

Deep past   Present

FIGURE 1.1 The continuum conceptualization

the most arcane subfields to be found at either end of the continuum, and the most applied to be in the middle. This is an oversimplification of our disciplinary relationships and the complexity of analysis within any individual field, but we have usually been prepared to be represented in this two-dimensional space. For example, in the School of Geosciences where I work, we represent our subject offerings to students along a continuum from geology at one end, physical geography in the middle and human geography at the other end. The so-called 'interface' subjects are situated, according to this dualistic logic, between the most human and the most physical.

There are a number of potential problems with these continua. The linear dualism of question framing tends to reproduce dualistic explanations and interpretations; for example, a focus on *either* the material *or* the symbolic dimensions of landscape, rather than a consideration of, say, the material expression of symbolism. It also tends to reinforce unprob-lematic definitions of and boundaries between humans and nature. Further, calls for interdisciplinary approaches then tend to be framed in terms that would lie in the centre; for example, the economics of the coal industry, or remote sensing of land clearance.

For the central subject matter of this book, I don't think this works well enough. I believe we have underestimated the creative interchange that can occur if we bend the whole thing over backwards. The recursive model (Figure 1.2a) thus brings together the 'opposite' ends of the continuum. This approach also brings together what many would consider the most arcane part of physical geography, the study of past environments, and the most arcane part of cultural geography, the construction of meaning about nature. They each need to be centre stage in this analysis. Recursiveness also brings together physical transformation and representation (Figure 1.2b), past and present, and human and natural agency (Figure 1.2c).

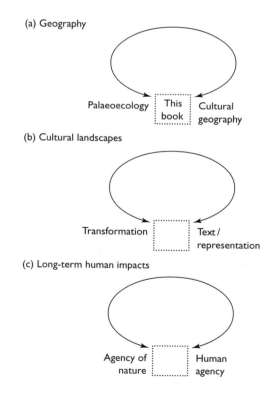

(a) Geography

Palaeoecology | This book | Cultural geography

(b) Cultural landscapes

Transformation | Text / representation

(c) Long-term human impacts

Agency of nature | Human agency

Figure 1.2 The recursive conceptualization

Now recursiveness on its own does not undo the dualism, it just recentres it somewhere else. To some extent this is simply an acknowledgement that the apparatus of Western thinking that we bring to bear is very deeply embedded. However, if we can also start to think multidimensionally, we can consider how any two ends or extremes pervade and interpenetrate each other. Most important here is to recognize the process of knowledge construction; the process of research and thinking (not to mention application to real-world environmental problems!) becomes part of the object of inquiry. Thus, for example, we can turn to textual readings of the story of human transformation of the Earth, looking critically at the ways in which we construct those stories (Figure 1.2b). We can acknowledge that nature has agency even in those contexts of apparent human domination, and that from the beginnings of human presence on the Earth people were constructing an understanding of the landscape and the human place in it (Figure 1.2c).

In being limited still by disciplinary structures I want to celebrate the achievements of the different traditions, being able to acknowledge their strengths and limitations. 'A geography that responds both to representations of the natural world and to biophysical process offers a richer toolbox for understanding the world' (Robbins, 1998: 83).

1.3 Contingent constructions

While distinctive contributions come from the different disciplinary traditions, there are important convergences in this subject matter between empirical and constructivist approaches. These centre around the ideas of *contingency* and *construction*.

Contingency, or the historical particularity of sets of circumstances, dominates recent thinking in palaeoecology, ecology and the historical humanities. It contrasts with deterministic approaches that interpret historical events, whether past or future, as the necessary outcome of certain processes. Application of appropriate analytical scales is important here. 'Big-picture' overviews often (unintentionally)

reinforce the idea that certain historical outcomes – the development of agriculture, the destruction of indigenous societies, land degradation – were inevitable. Not only does this reify categories, for example 'hunter-gatherer', which we should critically examine, it implies that the future is a foregone conclusion. In contrast, examining the details of how variable forces, processes and circumstances converged at different times and places in the past provides a more nuanced understanding, whether of forest composition or of European colonization of the Pacific. It also provides a more dynamic context in which to shape the future.

The idea of *construction* is one that has force in both the symbolic and material arenas. It provides for human agency in the biophysical and intellectual worlds. Although it has nuances of structures or buildings, it is used in the literature in much wider senses to discuss the ways in which people shape their landscape. It thus encompasses imaginative and symbolic domains as well. It also reminds us that the construction of knowledge, in this case about the past, is itself a process that needs explicit attention.

Under the umbrella of these two ideas, I draw out 11 theoretical underpinnings and seven operational guidelines which emerge from my interpretation of the most recent research on cultural landscapes and environmental change. They are used here to summarize the framework in which the book proceeds.

1.3.1 Theoretical underpinnings

1. The story of environmental change as derived from the scientific method is in the first instance a representation of events and processes that have happened, are happening or will happen. It is important to understand the processes by which science constructs its stories. However, contrary to some perspectives in the humanities, this does not mean that it is *only* a representation.

2. Biophysical processes have agency independent of humans, and humans are embedded in biophysical processes.

3. The definitions and boundaries of key categories in this analysis – human, nature, culture, landscape, environment, change – are socially constructed. There have been continuities and changes in those constructions over the full span of human history, and between different human groups.

4. In many contexts, then, it makes little sense, on either empirical or epistemological grounds, to talk of *separating* human and natural influences. They are mutually constitutive on both grounds.

5. Change is a much more normal state of affairs in the environment than stability, though both need to be considered at appropriate scales. This is now well established on the basis of palaeoecological and ecological evidence, as well as critiques of environmental stasis and equilibrium.

6. On the same basis, to understand humans as part of a biophysical system is to challenge the characterization of all human 'impacts' as implicitly damaging. It is not much help; we need to be more explicit about what is damaging and to what.

7. To emphasize the importance of mental constructions of landscape and nature is not to argue for a correlation between ethics and impacts. Communities with reverential attitudes to nature can have impacts that are at least as destructive as any other community (Tuan, 1968).

8. There are tensions and ambiguities here, and they can be creative: 'there are important understandings to be gained in seeking the social origins of instabilities and incoherences in our thoughts and practices – understandings that we cannot arrive at if we repress recognition of instabilities and tensions in our thought' (Harding, 1986: 243–4).

9. Diversity brings better science (and archaeology, geography and anthropology): 'The understanding afforded by divergent standpoints … is a crucial source of critical insight about what a dominant research community may be missing in its investigation of a given subject matter, historical or otherwise' (Wylie, 1995: 271).

10. Defined categories of human activity – for example, hunter-gatherer, agriculturalist, subsistence farmer – serve us only to a limited and generalized extent in understanding different types of interaction with the Earth. Deconstructive approaches and local-scale palaeoecological and ethnographic research point in the same direction here. There is considerable variability within categories, and unacknowledged shared traits between categories. We thus need to be vigilant about whether categorizations/labels/taxonomies are constraining our thinking in unconstructive ways.

11. For the same reasons, deterministic trajectories of change have little to recommend them. Contingency provides a better historical principle, one increasingly used in both the ecological and cultural literature: '"Contingency" is the label we will use that highlights the historical and local specificity that has to be accounted for in complete understanding and effective manipulation of ecological systems' (Parker and Pickett, 1997: 18).

1.3.2 Operational guidelines

1. For the purpose of this exercise, all landscapes are cultural, and all are understood as having multiple dimensions, including both material and symbolic.

2. Appropriate scales of analysis, in both space and time, are crucial.

3. The way in which we can use the long-term record of environmental change to assist with contemporary management issues is not at all self-evident. We cannot go backwards; we cannot just 'learn the lessons' of the past and easily effect transformation of our own social context.

4. Critical scrutiny should not be 'an abstract exercise in conceptual critique but should depend on a detailed (empirical) understanding of the conditions of knowledge production' (Wylie, 1995: 271). This is discussed in the context of Haraway's 'situated knowledge'.

5. We need to be vigilant about the ways explanations are built on biased bodies of evidence. For example, how does the dominance of the pollen record condition our thinking about past human interactions?

6. The empirical evidence can be as important as theory in deconstructing categorizations that have been taken for granted: for example, 'colonization', 'pristine baselines' and 'the Neolithic'.

7. While human activity is implicated in all Earth's systems, the limitations to our control are considerable.

1.4 Writing and reading this book

The research process, whether in a chemistry laboratory or an art gallery, is always a complex one involving interpretation, filtering and translation of knowledge and understanding. We tend to be more attuned to this process, and conscious of the partiality of our knowledge and voice, in a cross-cultural situation, such as when I as a white Australian write about Aboriginal lives. However, critical attention to the conditions of knowledge production is even more important in research contexts we are likely to take for granted, or regard as normal. In the context of this book, such contexts particularly include the production of knowledge about past and future environmental change.

It is no accident that many of the examples in the book come from former colonial corners such as Australasia and Africa. This work, both empirical and theoretical, is showing how the very structure of our knowledge and understanding about long-term human interactions with the environment is a product of particular historical circumstances arising out of the European colonial project. In biogeography, Stott (1998) summarizes this as the hegemony of forest ecology, the hegemony of equilibrium notions, and the hegemony of Europe and North America over the rest. Within archaeology, there are different traditions of inquiry on either side of the Atlantic.

How might we think differently about people, nature, environmental change and the past if we explicitly engage with the processes and legacy of colonialism, and if we give voice to indigenous and local understandings? The dominant message in this regard is one of humans within nature, as a force both creative and destructive, symbolically present and physically engaged, as expressed in underpinnings 3, 4, 6 and 10 (above).

This book is not an historical overview of human impacts on the Earth, or of cultural attitudes to the environment, although it provides signposts to important works in both those areas. There are big gaps in my coverage of topics, places and authors. It is not a comprehensive review of debates about 'culture' or 'landscape', although it does summarize the most relevant parts of these debates for issues of environmental change. The issue is not to define cultural landscapes, but to consider the multiple ways in which the concept has been used. The culture of writing about environmental change is just one of the cultures that demands critical analysis. Nor is this a set of cautionary tales or heroic myths from the past about what we should do in the future, although I do identify useful principles and processes.

Rather, it is an attempt to cut a particular path through a complex set of literature and build an argument about the conditions under which long-term environmental change is relevant to contemporary environmental issues. To this end, Part II will analyse the contribution of methodological and conceptual tools from the sciences. In Chapter 2 the focus will be on palaeoecology, and its particular contribution to our understanding of long-term human transformations of the landscape. With historical links to the work of George Perkins Marsh and Carl Sauer, cultural landscapes are understood in this literature as ones physically transformed by human action. However, the application of new techniques and expansion of palaeoecological work in Australasia, Africa and Asia is changing thinking on a number of themes. These include the influence of hunter-gatherers and the nature of change in forest ecosystems. Chapter 3 will consider how environmental change is understood within ecology and palaeoecology, focusing on the influence of so-called non-equilibrium ecology and its challenge to linear models of deterministic succession or circular ones of balance and stasis.

The following two chapters (Part III) constitute a section on the contributions of the humanities. In Chapter 4 the literature which deals with the social construction of nature and landscape will be reviewed. Multiple and shifting meanings of culture, nature and landscape, as seen particularly in the past three decades, will be considered. The notion of landscape within prehistoric archaeology will be discussed in most detail since it shows not only the changes within academic thinking, but also the long-term perspective of human perceptions of landscape. In Chapter 5 we shall consider how knowledge about environmental change is produced within Western science, using global climate change and deforestation as the major case studies.

The second half of the book (Part IV) will examine the application of these debates in selected contemporary environmental issues where a long-term perspective on human–environment interactions is particularly relevant. These issues are hard to isolate; many of the examples used would be applicable in several chapters. Chapter 6 will consider protected landscapes, including those explicitly defined by managers as 'cultural landscapes'. Examples of protection at international, national and local scales are used, with a focus on World Heritage cultural landscapes. In Chapter 7 we shall consider the concepts of preservation, restoration and creation in terms of both biotic and built landscapes. Management dilemmas here are implicitly temporal ones. Are we trying to return to the past, freeze the present, or create the future? Chapter 8 will attempt to come to grips with the human question; in what contexts do people have a place in landscapes? Chapter 9 will look at landscapes of identity, heritage and tourism, arguing that the process of valuing landscapes is a necessarily cultural process about which we should be as explicit as possible.

Part II

METHODOLOGICAL AND CONCEPTUAL TOOLS FROM THE SCIENCES

2

Transformed landscapes: human impacts and the palaeoecological record

2.0 Chapter summary

Within palaeoecological research, cultural landscapes are generally understood as those transformed by human action, a conceptualization particularly influenced by the work of Carl Sauer. Global overviews present evidence of accelerating and destructive human impacts over time. While in one sense this is an important and unarguable message, it should not be read in a deterministic way. Further, recent work at a range of scales shows considerable spatial and temporal variability. Three themes – colonization, hunter-gatherer/agricultural impacts and post-industrial changes – are reviewed. In each of these, rethinking of substantive and epistemological issues is under way, and constant methodological refinement is occurring. Vigorous debate over the timing and effects of human impacts in the South Pacific and elsewhere contributes to a rethinking of the concept of colonization. Similarly, clear demarcations between hunter-gatherer and agricultural impacts are difficult to maintain, and there is considerable variability within each of the categories. High resolution of the record of the past few hundred years in some parts of the world is facilitating a shift from inductive to deductive approaches within this field of research. Important issues within the three fields include the re-emergence of environmental agency, the importance of multiple palaeoecological proxies, and the influence on interpretation of spatial bias in concentrations of research.

2.1 Changing the face of the Earth

Interest in long-term human impacts on the landscape – or the Earth as transformed by human action – comes from several research directions. Palaeoecological researchers have for a number of decades recognized the possibility of anthropogenic signals in the records under study, although there has been considerable debate over their interpretation. Workers more interested in contemporary timescales and problems have also sought to place these in historical perspective. There are thus a number of overviews to which readers can be directed for more comprehensive coverage. A brief review of their themes and approaches helps to set the approaches of this book in context. It will also help us start to ask a central question: how does/should understanding of long-term impacts help us deal with contemporary environmental management issues?

The term 'cultural landscape' is used in this literature in a variety of ways, but in an overall sense that a cultural landscape is one physically transformed by human action. Temporal overviews thus present a story of increasing human impact over time, accelerating with recent population growth and technological change. Thus, for example, Simmons's (1989) *Changing the face of the Earth* uses a five-stage model of human history (Primitive man, Advanced hunters, Agriculture, Industrialists, the Nuclear Age) to chart human impacts. To help structure the mass of information, Simmons also uses the study of energy flow through ecosystems as the currency by which those impacts are measured. Goudie (1993) recognizes similar stages of cultural development, although he subsumes them within chapters structured by different elements of the biosphere.

In overviews which cover the whole of the Quaternary period (Mannion, 1991; Williams *et al.*, 1998), or focus on the Holocene – the past ten thousand years (Roberts, 1989) – the temporal structure provides a framework to understand increasing human impact towards the present, with varying amounts of detail depending on the intended audience. Most such overviews acknowledge that the impacts go in two directions: of environmental change on people, and of people on environment (e.g. Bell and Walker, 1992; Chambers, 1994).

Second, there are overviews of our present dilemmas, which usually include some temporal context in their introductory chapters. For example, Meyer (1996) refers to the formula by which human impact can be thought of as population multiplied by affluence (demand on Earth's resources per person) multiplied by technology. He then provides an historical review of increases in each of the three elements.

Not all these authors use the term 'cultural landscape', and where they do, the meaning is often implied rather than explained. It is useful, then, to focus on the traditions in which these authors see themselves as working – to whose influence do they defer? The intellectual influences most frequently acknowledged are those of George Perkins Marsh, Clarence Glacken, W.L. Thomas and Carl Sauer. In a review that ends well before contemporary interest, Glacken (1967) showed that the questions were not new. For him, the question of human transformations of the Earth is one of the three most persistent questions that 'men' have asked about their relationship to the habitable Earth since the beginning of Western thought. Thomas's 1956 edited volume *Man's role in changing the face of the earth* is often seen to signal a turning point in contemporary awareness, at least by the academic community, and its lineage is traced to Marsh's 1864 work *Man and nature; or, Physical geography as modified by human action* (Marsh, 1965 [1864]). Later works acknowledge these influences in

their titles and forewords, e.g. Simmons (1989); Turner *et al.*, 1990).

In the specific context of human impacts writing, the term 'cultural landscape' is most often associated with the work of Carl Sauer. (Although Sauer is generally considered responsible for the transfer of the concept into English, Jones and Daugstad (1997) attribute its first use to the German geographer Friedrich Ratzel (1895–6).) For Sauer,

> The works of man express themselves in the cultural landscape. There may be a succession of these landscapes with a succession of cultures. They are derived in each case from the natural landscape, man expressing his place in nature as a distinct agent of modification. Of especial significance is that climax of culture which we call civilization. The cultural landscape then is subject to change either by the development of a culture or by a replacement of cultures. The datum line from which change is measured is the natural condition of the landscape. The division of forms into natural and cultural is the necessary basis for determining the real importance and character of man's activity. …
>
> The natural landscape is being subjected to transformation at the hands of man, the last and for us the most important morphologic factor. By his cultures he makes use of the natural forms, in many cases alters them, in some destroys them. …
>
> The cultural landscape is fashioned from a natural landscape by a culture group. Culture is the agent, the natural area is the medium, the cultural landscape the result. (Sauer, 1965 [1925]: 333, 341, 343)) (Figure 2.1)

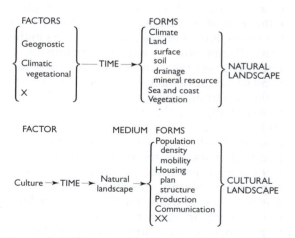

FIGURE 2.1 The natural and cultural landscapes, as conceptualized by Sauer (1925)

The main features of Sauer's 1925 conceptualization are reflected in much of the literature referred to in this chapter: an unproblematic separation of the natural and the cultural; the idea of a pre-human baseline; humans as an agent of transformation; and the influence of varied causes operating over time. The so-called Sauerian view of culture has been subject to critique in the past few years, a debate that is reviewed in more detail in Chapter 4. However, it is important to remember that in the middle decades of the twentieth century Sauer and his colleagues were still having to address and combat ideas of environmental determinism, whereby particular environments were argued to give rise to particular cultures. They have done that so successfully that some writers today argue that we have lost sight of the agency of the environment, and of natural-scale changes (Nunn, 1999).

The tools to reconstruct cultural landscapes over remote times and places have changed dramatically in recent decades. Using a great variety of palaeoecological indicators, set in a temporal context with radiometric dating techniques, we can review human impacts over longer timespans and with better spatial resolution (Box 2.1). Just as Sauer's ability to get at these questions was limited by the tools available – radiocarbon dating, for example, was not available until right at the end of his life – it is important to identify the ways in which our current thinking is a function of our tools. The literature is heavily weighted towards human impacts on vegetation, particularly in terms of the earliest evidence. But as examples discussed below show, the pollen record as often studied (or studied for other purposes) can bias thinking about human impacts in particular ways. These include the scale of change, the expectation of deforestation, and overemphasis on introductions of palynologically visible taxa.

In such a huge and diverse field of interest, overviews such as those referred to above provide essential perspective for students and researchers in various fields. In one sense a story of increasing and accelerating human impact is an unarguable and important message (Table 2.1). So too is the understanding that these impacts have occurred over a period of great climatic change with asso-

---

Box 2.1   Anthropogenic indicators in sedimentary records: some examples

Pollen

- Particular indicators need to be ecologically appropriate:
  (a) *Old World* – pollen of cultivated plants in a variety of combinations, and associated disturbance indicators (Behre, 1981).
  (b) *New Zealand* – decline in forest pollen, increase in *Pteridium* (McGlone, 1983).
  (c) *Southeast Asia* – increase in abundance of secondary forest tree taxa, e.g. *Trema, Macaranga, Dodonaea*, and increase in herbs (Flenley, 1988); problems and potentials in identification of cultivar pollen and using weeds as indicators (Maloney, 1994).
  (d) *Africa* – some forest increases can be anthropogenic (Taylor *et al.*, 1999).
  (e) *New Guinea* – high proportions of degraded *Nothofagus* pollen, grass increases if accompanied by forest decline and swamp grasses excluded, *Casuarina* increases as part of agro-forestry, *Pandanus*. Pollen of many known crops, e.g. *Ipomoea batatas* and *Colocasia esculenta*, are not visible in the fossil record (Haberle, 1994).

Diatoms

- Indicate eutrophic conditions at Lake Coba, Mexico, at a time when there is no support from other indicators for a climatic explanation. Diking for a reservoir is argued to be a viable hypothesis (Leyden *et al.*, 1998: 119).

Microscopic charcoal

- A record of fire–rainfall events (Clark, 1983).
- Needs particular attention to scale (Clark, 1983).

Influx of inorganic sediments

- As measured by loss-on-ignition (Flenley, 1994).
- Magnetic mineral analysis (Haberle, 1994).

Charcoal and wood

- Human impacts have been identified in highland New Guinea as early as 28,000 BP, on the basis of charcoal in slopewash deposits in the Baliem Valley, Irian Jaya (Haberle *et al.*, 1991). Although the connection to humans is circumstantial, natural fire is extremely rare in these cool, wet montane forests (*see also* McGlone, 1983).

Soil instability

- Soil instability can obviously have other causes, but when associated with charcoal is extra evidence (McGlone, 1983).

Phytoliths (plant opal silica bodies)

- Identification of major New World domesticates, e.g. maize, squash, gourd, and many utilized wild plants, e.g. palms and bamboo. But not New World root crops such as manioc, potato and sweet potato (Pearsall, 1994). Future potential considerable since many palynologically invisible taxa produce phytoliths.

Plant macrofossils

- Native ruderals (disturbed ground plants) and introduced weeds (Baker *et al.*, 1993b).

Molluscs

- Particular species, species associations and diversity as indicators of vegetation interference and clearance on the chalklands (Evans, 1994).
- Land snails as transported through Pacific islands by people (Kirch, 1984: 136–7).

Insects

- Beetles associated with dung, polluted waters and cultivated plants as evidence of European settlement in North America (Baker *et al.*, 1993).

ciated shifts in sea levels, vegetation formations and landscape alteration. Specialist or regional syntheses (e.g. Birks *et al.*, 1988; Bottema *et al.*, 1990), have had greater scope to explore the complexities of different scales of analysis, or different ecological contexts. However, there are a number of shared features of the big picture:

- Dualistic views of humans and environment are prevalent.
- Humans are seen as disturbers of the natural order.
- Humans have impacts on, and progressively override, the forces of nature as they increase their ecological dominance.
- This results in the gradual enculturation of environments.

- Prehistoric change is most strongly associated with agriculture, in the mid-Holocene.

For example, in one of the most detailed overviews, Roberts (1989) shows explicit links between the shift from hunter-fisher-gatherer to agriculture to urban-industrial systems (Figure 2.2). Cultural landscapes, associated with modification of the environment, appear with agro-ecosystems, and are the period of the Holocene when human influences (H) are seen to override environmental ones (E).

There is a risk that historical trajectories will be read in a deterministic way that forecloses other possibilities (in both interpretation of the past and implications for the future), although this is certainly not the intention of the authors

TABLE 2.1 Selected forms of human-induced transformation of environmental components: chronologies of change

A Quartiles of change from 10000 BP to the mid-1980s

Form of transformation	Dates of quartiles[a]		
	25%	50%	75%
Deforested area	1700	1850	1915
Terrestrial vertebrate diversity[b]	1790	1880	1910
Water withdrawals[c]	1925	1955	1975
Population size	1850	1950	1970
Carbon releases[d]	1815	1920	1960
Sulphur releases[e]	1940	1960	1970
Phosphorus releases[f]	1955	1975	1980
Nitrogen releases[d]	1970	1975	1980
Lead releases[d]	1920	1950	1965
Carbon tetrachloride production[d]	1950	1960	1970

B Percentage of change by the times of Marsh and of the Princeton symposium

Form of transformation	Percentage Change	
	1860	1950
Deforested area	50	90
Terrestrial vertebrate diversity[b]	25–50	75–100
Water withdrawals[c]	15	40
Population size	30	50
Carbon releases[d]	30	65
Sulphur releases[e]	5	40
Phosphorus releases[f]	<1	20
Nitrogen releases[d]	<1	5
Lead releases[d]	5	50
Carbon tetrachloride production[d]	0	25

Source: Turner et al. (1990: Table 1.3)

[a] Calculations assume a baseline or pristine biosphere about 10000 BP and 100 per cent change as of the mid-1980s. Percentages refer to the total of the later or 100 per cent figure

[b] Number of vertebrate species that have become extinct through human action since 1600. Does not include possible waves of Pleistocene and earlier Holocene human-induced extinctions because of continuing controversy over their nature and magnitude

[c] Total amount of water now withdrawn annually for human use

[d] Total mass mobilized by human activity

[e] Present human contributions to the sulphur budget

[f] Amount of phosphorus mined as phosphate rock

cited above. The dangers of a teleological approach are recognized by Simmons (1993a: 110). Nevertheless, practising and training environmental scientists are less likely to go from the overviews to the source literature where more detailed discussions of scale, complexity and interacting causal processes are to be found.

Further, most overviews are dominated by northern-hemisphere, particularly north-west European, perspectives. The attempt here to increase the visibility of southern-hemisphere work not only starts to fill the gaps from an empirical point of view, but also shows that the nature and limitations of the evidence itself continue to influence our understanding and

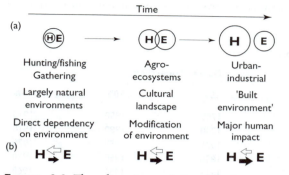

FIGURE 2.2 The changing relationship between humans (H) and the natural environment (E) over the course of the Holocene, including (a) nature of interaction, and (b) relative impact
Source: Roberts (1989: 184). By permission of Blackwell Publishers

the terms in which debates are conducted. In a global overview of the earliest palynological evidence of human impact on vegetation, Walker and Singh (1994) show that only in the Australasian/South-East Asian region are impacts suggested to pre-date the Holocene. The exception to this is Hoxne, South-East England, where last-interglacial vegetation disturbance has been attributed to human activity (West and McBurney, 1954; West, 1956). In North America even agricultural impacts have low pollen visibility, showing up in less than 20 of over 300 published pollen diagrams spanning the past 12,000 years (McAndrews, 1988), although site selection and sampling have not always been appropriate for seeing human impacts (Delcourt, 1987). The studies also emphasize the importance of understanding local ecological conditions, particularly in terms of interpreting fire and charcoal records.

As a complement to the overviews cited above, three slightly different areas of research are reviewed in this chapter. I hope to convey the excitement, innovation and relevance of recent research. I focus on regional syntheses which bring together data from a number of sites. Additional themes which this work brings include:

- Increased emphasis on potentially early hunter-gatherer impacts. There is much controversy in this field, and few areas of

consensus. Disagreements have, however, led to more explicit attempts to delineate criteria of human presence and activity.
- These methodological debates have fed back into less controversial areas, such as agricultural impacts. At different scales, very clear demarcations between hunter-gatherer and agricultural impacts disappear, and each category of land use shows considerable internal variability.
- There has been a recent reappearance in the literature of the agency of the environment, and attempts to discuss this in a non-deterministic way.
- Each of these points connects with debates over the mutual constitution of society and environment that will be discussed in Chapter 4.

2.2 Landscapes of colonization

In debates over human transformations of the Earth, writers have traditionally distinguished between old lands such as Africa, where humans coevolved with other parts of the biosphere, and newly colonized parts of the world. Human colonization of new lands theoretically provides the laboratory experiment about human transformations of those environments – they provide the 'datum line', in Sauer's terminology. If a recent argument that African hunters were influencing biospheric processes, including climate, as early as 3.5 million years ago (Burchard, 1998) is correct, even this theoretical baseline may disappear, but we will maintain it for the purposes of the present discussion.

What sorts of transformations occur when people enter an ecosystem which has not experienced them before? To restate the obvious, people always come from somewhere, bringing with them plants, animals and cultural concepts. Drawing on Anderson (1952), Kirch (1984: 135–9) referred to these as 'transported landscapes'. In different contexts the outcomes include extinctions and introductions of biota, and both deliberate and adventitious impacts on pre-existing vegetation and landforms. Two examples with very different timescales of colonization, Australia and New Zealand, show that this debate is never straightforward. It is important to emphasize that the issues considered in this section are particularly controversial; it is not my intention to try to solve them. Rather, I focus on how the evidence is used, and the links between transformation and colonization. There are thus important parallels between these debates and those in other parts of the world, particularly the Americas.

For more than twenty years, archaeological debates about the timing of first human occupation of Greater Australia (present-day Australia and New Guinea) have been intertwined with palaeoecological ones about the mooted impacts of those first settlers on both flora and fauna (Singh *et al.*, 1981; Horton, 1982). When did people arrive, did they change the vegetation through burning it, and did they cause the extinction of the megafauna? An important dimension of the controversy is that the timing of Aboriginal burning as interpreted from the pollen and charcoal record has always been much earlier than the archaeological evidence, even though both have changed over the decades. For example, Aboriginal burning was suggested at Lake George in south-eastern Australia at around 125 ka (Singh *et al.*, 1981) at a time when the archaeological dates were still around 35–40 ka.

Current archaeological estimates for the time of human arrival vary from around 40 ka (O'Connell and Allen, 1998; Mulvaney and Kamminga, 1999) to around 60 ka (Roberts *et al.*, 1990, 1994) or beyond (Fullagar *et al.*, 1996; Roberts *et al.*, 1998; Thorne *et al.*, 1999). Archaeological debates centre on the relative efficacies of dating techniques (radiocarbon, thermoluminescence and optically stimulated luminescence), and on taphonomic issues such as post-depositional movement of artefacts within sediments. The latter is relevant because the sediments rather than the contained artefacts are being dated.

Even on the shortest scenario, then, the time frame covers a period of significant environmental change. It has thus been recognized for some time that palaeoecological separation of anthropogenic and climatic influences requires records which cover at least the last glacial–interglacial cycle and preferably longer

(Kershaw, 1986). A recent overview of nine such records (Kershaw *et al.*, in press) spans south-eastern Australia to the Indonesian offshore region, and examines charcoal within 'time slices selected to represent major identified past climatic phases' (Figure 2.3). Charcoal peaks are present in drier phases (oxygen isotope stages 2, 4 and 6) and during times of major climate change. In addition there is at least one charcoal peak in all records during the period 30–40 ka, one of the most climatically stable periods of the late Pleistocene. The burning had variable impacts on vegetation. At Lynchs Crater in north Queensland, drier araucarian rainforest began to be replaced by eucalypt woodland, and in the Banda Sea impacts on both Indonesia (reduction of Dipterocarpaceae rainforest) and northern Australia (expanded grasslands relative to eucalypts) are indicated. In the southern areas the vegetation impacts were less pronounced. The authors attribute these peaks to Aboriginal burning – which they infer 'impacted the whole of the Australian landscape, as well as some parts of Indonesia, within a short period of time' (p. 12) – but argue that the effect was to accelerate existing trends within vegetation rather than to transform whole landscapes.

Kershaw *et al.* (in press) now consider that this data set provides some support for the model of late colonization, about 40 ka, since there are no charcoal peaks in the 50–60 ka age range. Charcoal peaks with associated vegetation changes around 135 ka at several sites are no longer attributed to human agency (*see* Kershaw *et al.*, 1993), and vegetation changes at around 175 ka at two of the northern sites do not have associated charcoal peaks. In these earlier events, 'vegetation change appears to have preceded charcoal peaks' (p. 11), and it is suggested that fire is a result of climatically driven drift towards more open vegetation. Thus while fire activity is interpreted to have increased throughout the period of these records, 'late Quaternary vegetation changes have been less dramatic than originally suggested' (p. 11). This evidence is in accord with geomorphological evidence for increasing aridity over the past few glacial–interglacial cycles.

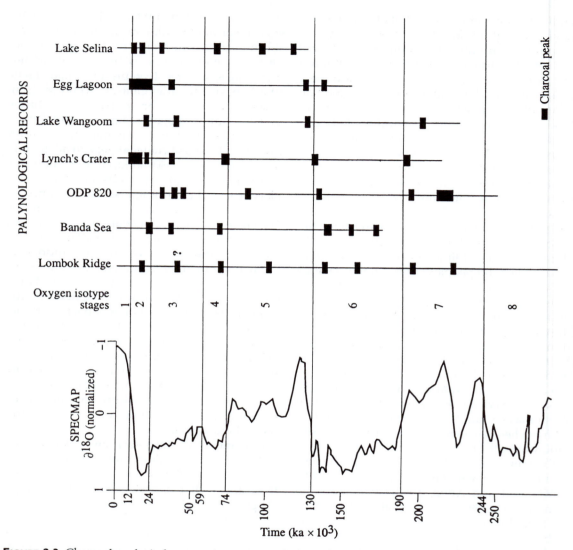

FIGURE 2.3 Charcoal peaks in long continuous records from the Australian region in relation to the marine oxygen isotope (SPECMAP) record
Source: Kershaw *et al.* (in press). By permission of Peter Kershaw

The question of landscape transformation extends beyond the vegetation to the fauna, and particularly to a suite of larger fauna which became extinct in Greater Australia at a variety of unknown times within the past 100,000 years. The relative influences of climate and people have been debated for more than a century; the most recent advocacy of a climatic explanation comes from Horton (2000), and of overkill by people from Flannery (1990, 1994).

2.2.1 New Zealand

New Zealand is an example of the islands of the south and west Pacific that illustrates the transformation of isolated islands by entering peoples, although it is much larger than most and lies within temperate latitudes. As this case study shows, there are parallels with Australia in that debates about human impact are closely entwined with those about timing of colonization. The transformation of the islands was influenced by four main processes: introduction of new plants and animals; clearance of native vegetation for agriculture; hunting of native fauna for food; and accelerated erosion of hillsides and consequent lowland deposition (Enright and Gosden, 1992).

There are two competing narratives in this literature. The dominant one is that human impacts on islands were sudden and dramatic. This is clear from, for example, Fiji within the past 3000 years (Hope, 1999a), Yap (federated states of Micronesia) about 3300 BP (Dodson and Intoh, 1999), Mangaia (southern Cook Islands) at 1600 BP (Kirch and Ellison, 1994), and Easter Island around 630 BP (Flenley, 1994). It is important to point out that there is an alternative narrative (Nunn, 1994, 1999) which argues that human influences have been overstated relative to others.

In the New Zealand context the mooted human impacts include anthropogenic deforestation, extinction of the moas and other birds (Illustrations 2.1a and 2.1b), and decline in seal and fish populations (Anderson and McGlone, 1992). As in Australia, this is discussed in the context of a debate over whether New Zealand has a short (<600 yr), intermediate (*c.* 1000 yr) or long (>1500 yr) prehistory (Sutton, 1987; Holdaway, 1996; McGlone and Wilmshurst, 1999). And, as in Australia, the question focuses on whether palaeoecological evidence of vegetation and landscape disturbance provides

evidence of human impacts independent of the archaeological record. Whereas in Australia the dating debates of relevance centre on radiocarbon compared with luminescence techniques, in New Zealand the questions are over the contamination and precision of radiocarbon dating (e.g. Newnham *et al.*, 1998a; McGlone and Wilmshurst, 1999).

Recent resolution of the date of the stratigraphically critical Kaharoa Tephra at 665 ± 15 BP (*c.* 600 cal BP – calibrated radiocarbon years before present) (Lowe *et al.*, 1998; Newnham *et al.*, 1998b) clarifies these issues. In an analysis of 11 pollen records containing the tephra, Newnham *et al.* (1998) argue that inferred human impacts occur around the time of tephra deposition or later, lending support to the late colonization model. Others argue that early Polynesian impacts cannot be ruled out as explanations for early fire and disturbance, such as that around 1800 yr BP on Great Barrier Island (Horrocks *et al.*, 2000). However, this possible early colonization period is also acknowledged to coincide with a period of increasing drought frequency which would have affected vulnerability to fire.

Most of the New Zealand flora is not fire adapted, and it is usually assumed that this factor and long successional recovery periods for forest made the vegetation particularly vulnerable to the fires of early colonists (Ogden *et al.*, 1998). Reviewing dates from wood charcoal in South Island soils, and microscopic

(a)

ILLUSTRATION 2.1a Excavation in progress, moa-hunting site, Shag River mouth, New Zealand (by permission of Atholl Anderson)

charcoal from North Island sedimentary cores, Ogden *et al.* (1998) show that fire has existed throughout the Holocene, with apparent increases in frequency after about 3 ka in the South Island and 7 ka in the North, both well before human arrival. Once again, increased climatic variability, possibly associated with enhanced El Niño – Southern Oscillation (ENSO) activity, is implicated. There is also a problematic peak in soil charcoal about 1000 BP, within the contested period. Ogden *et al.* distinguish between 'first anthropogenic fires', which will not be distinguishable from the background level of natural burning, and 'significant human impact', which will (Ogden *et al.*, 1998: 694; Illustration 2.2).

2.2.2 Palaeoecological criteria of human presence and impacts

Although none of these debates is resolved, the excitement generated by the research has stimulated the development and application of new techniques and strategies, some examples of which have been mentioned. Some of the technical advances are explored in more detail in other volumes in this series; I discuss here the broader question of how research design constrains and advances thinking. A diverse set of strategies have been proposed to examine the impacts of human activities (*see* Box 2.2 for a summary of examples). For the most part, these do not constitute testable hypotheses in

ILLUSTRATION 2.2 Bracken, tussock grassland and forest remnants, Port Hills, Canterbury, South Island, New Zealand. This landscape is interpreted by McGlone to be the result of forest clearance by early Maori (by permission of M.S. McGlone)

Box 2.2 Strategies and palaeoecological site selection for identification of human impacts: some examples

- Small pollen sites – more likely to reflect small-scale human activities (Brown, 1999).
- Separation of pollen studies from more explicit climatic proxies, e.g. collapse of Classic Maya civilization (Hodell *et al.*, 1995).
- Separate natural and anthropogenic fire frequencies – calculate return interval (Ogden *et al.*, 1998).
- Predict conditions of anthropogenic invisibility (even though present) (Barham, 1999; Head, 1996).
- Synchroneity of disturbance – varies with ecological conditions and occupation history. For example, rapidity of change in New Zealand is seen to support human factors (McGlone, 1983); non-synchroneity is argued by Haberle (1993) for the humid aseasonal tropics.
- Compatibility between archaeological and palaeoecological evidence – most workers in Australia would argue for this (see Head (1994a) for review). Not seen as necessary in remote island context, even among people with internal disagreements (Kirch and Ellison, 1994; Spriggs and Anderson, 1993).
- Examine cessation of traditional burning after colonization.
- Compare adjacent mainland and island sites with different occupation histories (Hope, 1999b).
- Multi-method – post-European Iowa (Baker *et al.*, 1993b).

Box 2.3 Scenarios and hypotheses about human presence and impacts in Central America and northern South America, 14–8 ka BP (from Cooke, 1998)

A human presence in Central America, Colombia and Venezuela (outside 'Amazonia') before the Late Glacial Stage (LGS)(~14–10 ka) requires substantiation. A survey of recent archaeological and palaeoecological evidence allows the specification of inferred high-probability scenarios, alternate scenarios whose resolution awaits better-quality data, and areas of investigation whose encouragement would improve knowledge.

High-probability scenarios

- A small and widely dispersed pre-Younger Dryas (El Abra Stadial, ~11–10 ka) human population.
- The pre-Younger Dryas (YD) consumption of extinct herbivores, including horse and gomphotheres, by cultural groups who made bifacial lanceolate points and carefully trimmed scrapers.
- More or less coincident with the YD and El Abra Stadial (10.5 ± 0.6 ka), a widely distributed, culturally affiliated and highly mobile ('Palaeoindian') population with a distinctive stone tool kit.
- The use of different forest and open vegetation types throughout the LGS from sea level up to *c.* 2500 m (except non-seasonal lowland rainforest).
- At or soon after the LGS/Holocene boundary, widespread forest-dwelling settlements, whose inhabitants habitually collected forest plant products, some of which appear to have been under domestication by 8 ka.

Alternative scenarios which cannot be confirmed by current data

- Humans were/were not present during the upper Pleniglacial and did/did not use extinct Pleistocene mammals such as mastodons.

- All pre-El Abra Stadial populations were/were not acquainted with bifacial reduction processes.
- Pre-Holocene populations did/did not habitually collect wild plant foods.
- The extinction of Pleistocene mammals did/did not coincide chronologically with the disappearance of fluted points.
- Extant fauna, such as white-tailed deer, were/were not an important LGS hunting resource.
- LGS and early Holocene tool kits represent *either* short-lived, but areally distributed artefact types, *or* coeval assemblages belonging to different cultural groups

with distinct subsistence orientations.
- Hunter-gatherers using bifacial projectile points and large blade/flake scrapers did/did not survive after 8 ka coevally with incipient cultivators.
- Universal forest expansion and/or increased sedentism after climatic amelioration did/did not lead to the development of several economic systems strongly correlated with specific geographic regions and/or vegetation types (i.e. an 'Andean hunting tradition', 'a Central American pre-montane horticultural system').

the sense discussed in Section 2.4 (p. 30). They depend heavily on issues of temporal and spatial correlation between, for example, palaeoecological and cultural indicators. However, increased methodological resolution is facilitating more explicit approaches (*see* Box 2.3 for an example).

The specific nature of landscape transformation feeds back into a rethinking of the concept of colonization. It is not just a question of whether people were there or not, but what they were doing and in what ecological contexts. It is interesting that much of this evidence is coming from islands, often considered the archetypal laboratories of instantaneous destructive impact. In the New Zealand context we can, then, distinguish between:

- Colonization as discovery – early exploratory or accidental landings, possibly of a single canoe, from which rats escaped. This did not result in prolonged or widespread human occupation, but resulted in impacts on small ground-dwelling birds on which the rats preyed.
- Systematic colonization – later near-simultaneous archaeologically visible occupation associated with rapid, widespread deforestation and moa extinctions.

Detailed multi-method research on cave deposits in north-west Madagascar is clarifying the mechanisms of faunal extinction there

within the past 2000 years (Burney *et al.*, 1997). Humans have long been the main suspects in an event that was, in geological terms, extremely rapid. Many aspects of the extinction process are still unclear; there is, for example, no definite evidence for direct interaction between humans and extinct species other than tortoises. However, Burney *et al.*'s research shows that fire was an important environmental variable long before people arrived. Some extinct animals may have persisted for at least a thousand years after human arrival on Madagascar, probably owing to the remoteness from human activity of the north-west coast.

None of these examples denies the profound impacts of colonizing peoples. However, if the propensity for ecosystemic destruction is seen as part of an essential human nature, then the past is of little use because the outcomes are inevitable. Alternatively, if impacts are contingent on the conjunction of particular circumstances, the story of the past can be useful for contemporary management. In this case clarification of the rates and precise mechanisms of change becomes an important contribution.

2.3 Hunter-gatherer/agricultural landscapes

Hunter-gatherers have long been distinguished from agriculturalists partly on the basis of attributes they lack, attributes that relate to interactions with landscape. Thus, it is argued, hunter-gatherers do not plant or tend gardens, they do not substantially modify their environment, nor do they purposefully manage it. In nineteenth-century parlance, they lack purchase on the land. Similarly, in some circles agriculturalists are defined by the changes they make to the genetic make-up of plants or animals. These arguments have come under assault from anthropological, archaeological and palaeoecological research in recent decades, to such an extent that clear delineation between 'hunter-gatherers' and 'agriculturalists' is rejected by some. It is beyond the scope of this book to review these debates, some of which are discussed in Chapter 4, but one important point emerges. Clear assessment of hunter-gatherer and agricultural impacts requires us to suspend judgement about the pre-existence of those categories. If they cannot be distinguished on landscape grounds, then this hitherto fundamental human taxonomy is further challenged.

This is not to deny the profound differences between some hunter-gatherer and some agricultural interactions with the environment. Rather it is to assess the evidence for landscape transformation with an open mind on whether we can 'define a rubicon with social consequence and substance that can systematically differentiate [hunter-gatherer from small-scale agricultural or pastoral] societies' (Feit, 1994: 438). Thus in this section I do not distinguish between hunter-gatherer and agricultural impacts, but instead review some case studies which illustrate methodological debates about the types of impacts/transformations which are visible in the palaeoecological record. Further, I want to show how those debates feed back into others relating to hunter-gatherer/agricultural transitions, and challenge us to rethink assumed categories.

2.3.1 North-west Europe

Mooted human transformations of the landscape of north-west Europe include assistance of rapid early-Holocene migration of *Corylus* (hazel) (Huntley, 1994), Mesolithic defor-

estation associated with fire (Behre, 1988; Simmons, 1994; Wiltshire and Edwards, 1994) and enhanced development of blanket mire (Caseldine and Hatton, 1994; Moore, 1994).

Brown (1997a) provides a critique of what he sees as this traditional account of 'clearance' or 'deforestation' in the Mesolithic/Neolithic of Europe, whereby an overall decline in tree pollen from around 3000 BC onwards in the British Isles is widely interpreted in terms of deliberate clearance and land management. He argues that

> the pollen evidence amounts to a record of decreases in canopy-forming taxa and the existence of open areas, generally followed by a delayed regeneration of woodland in the Mesolithic but an increasingly common lack of regeneration in the Neolithic and after. (p. 138)

Since there is a lack of archaeological evidence for ringbarking or stripping of trees, and no clear evidence of the use of mature trees on a large scale until the mid- to late Neolithic, Brown favours an opportunistic deforestation model, whereby a variety of natural factors (including drought or windthrow) are the initial causes of deforestation, with humans being the primary factor preventing regeneration.

A couple of points are of interest here. One is Brown's call for 'a more fragmentary narrative' of human impacts, in which natural events and ecological instability have an increased role. Second, his interpretations, while they have theoretical ramifications, depend in the first instance on questions of methodology. Thus Brown (1999) calls strongly for more localized pollen analyses, and for recognition of the mismatch between palynological and archaeological sampling methodologies. Palynologists, he argues, have generally preferred sites which will provide regional vegetation histories. (Early palynology, keen to establish the field as a source of palaeoclimatic information, deliberately tried to exclude local variability.) Sites with larger pollen catchments may suggest vegetation homogeneity when the reality is much more patchy. There is also a bias in Britain towards upland and bog sites.

Alluvial sites, Brown argues, provide the means to overcome both these problems. They are inherently local sites, but were also important foci for prehistoric occupation, particularly in regard to the watering of cattle. He illustrates this with a study of the Soar and Nene valleys, in England's east Midlands, in which palynological and archaeological evidence was examined at comparable (local) scales (Brown, 1999). Small-scale localized disturbances, in which natural events such as flooding and windthrow were important, occurred in the Neolithic, but these were temporary. Larger-scale deforestation with considerable spatial variability occurred in the Bronze Age, and by the mid- to late Iron Age the floodplains were almost totally deforested for stock management, with some cultivation on the terraces and fringes (*see also* Willis and Bennett, 1994, for a related discussion of scale in the Balkans).

In turn, these findings stimulated questions about human perceptions of the changes. Was the late Neolithic/early Bronze Age vegetation change associated with ritualization of the floodplain as recorded by barrow construction? And what was the cultural significance of the shift between the Bronze and Iron ages from a woodland landscape with openings to an open landscape with woodlands (Brown, 1999: 7)? These issues are discussed further in Chapter 4.

2.3.2   Central Africa

Even though both a long period of human occupation and the background variability of early Holocene landscapes are acknowledged, most workers in north-west Europe consider that early Holocene forests are, for all practical purposes, 'pristine' landscapes in terms of assessing human impacts. The issue is more problematic in parts of the world with longer human histories, for example the interlacustrine region of Central Africa – although, as Taylor *et al.* (1999) show, methodological issues such as scale of analysis, catchment size of pollen sites and articulation with the archaeological evidence are just as important. Further, establishment of independent climatic proxies, where possible, assists in teasing apart the human signals. For five locations in western Uganda over the past three thousand years, Taylor *et al.* use the following indirect proxies of human activity: reduced organic matter; increased sedimentation rate; charcoal; pollen from taxa associated with primary (*Olea*) and secondary (*Celtis* and *Trema*) forest; degraded soils (*Dodonaea*); and open vegetation (*Vernonia* and *Poaceae*) (Figure 2.4).

Two phases of forest disturbance are identified. The first, at Muchoya and Kabata Swamps and at Lake Victoria around 2200–2000 BP, could reflect the introduction of

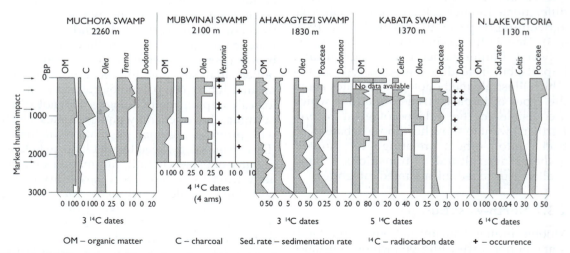

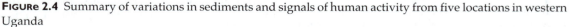

Figure 2.4 Summary of variations in sediments and signals of human activity from five locations in western Uganda
Source: Taylor *et al.* (1999). By permission of David Taylor

iron working and agriculture to the region. The more widespread phase, associated with charcoal peaks with open and disturbance vegetation indicators at 1000–800 BP, correlates with Later Iron Age technological and settlement changes. However, the climatic context in which these changes were occurring was not static: conditions became more humid at this time, perhaps promoting major population movements into the area. Nor does climatic influence operate only in one direction: Kabata and Lake Victoria show a third phase of forest disturbance at 600–400 BP, a time of relatively dry climates. Mubwinai Swamp shows disturbance within the past 200 years, possibly associated with the colonial period.

In many places that would be considered a fairly comprehensive set of indicators. However, Taylor *et al.* argue that such a suite of proxies, although commonly used in the tropics, unnecessarily constrains our understanding of human impacts:

> the choice of evidence has been partly guided by the 'orthodoxy' of deforestation, which views human activity in general as ecologically disruptive, intrusive and negative. Thus it is usually assumed that the onset of agriculture is, in the absence of direct indicators such as pollen from domesticated plants, marked by evidence of forest clearance and burning. (1999: 7)

Thus African evidence for variable agricultural strategies is ignored. For example, 'shifting cultivators often make use of natural forest gaps and ... "forest" can be promoted by certain forms of agriculture, such as low intensity cultivation of oil palm (*Elaeis guineensis*)' (1999: 7). Also problematic in the assumption is the role of fire and change in natural vegetation dynamics, with pollen samples dating back to the last glacial showing a long period of fire history.

Although they see the human signal in these five sedimentary records for the late Holocene as the loudest one, and thus human activity as the major cause of change, Taylor *et al.* thus emphasize that this interpretation is limited by the evidence. This does not demonstrate that these major and disruptive human impacts were the only form of human activity influencing ecosystems in the past three thousand years. Long-term coexistence of 'regenerating' and 'stable' patches of forest is indicated at Ahakagyezi Swamp (Taylor, 1993).

2.3.3 New Guinea and Indonesia

Human impacts have been identified in highland New Guinea as early as 28,000 BP, on the basis of charcoal in slopewash deposits in the Baliem Valley, Irian Jaya (Haberle *et al.*, 1991) (Figure 2.5). Although the connection to humans is circumstantial, natural fire is extremely rare in these cool, wet montane forests. The relevance of understanding the long-term role of fire in tropical forests was further highlighted by the widespread forest burning across Malesia during the El Niño event of 1997–98. Haberle *et al.* (in press) analysed 10 sites from the eastern Indonesian archipelago and Papua New Guinea for which consistent charcoal measurements and good chronological control since 20,000 BP were available. These were compared with independent climate proxies such as ocean sediment records and glacial change, and with archaeological data on human occupation.

At this regional scale two main periods of high fire frequency were identified: during the last glacial transition (17–9 ka) and to the mid- to late Holocene (past 5 ka) (Figure 2.6). Although people were in the region throughout the period, Haberle *et al.* suggest that fire frequencies were related more to climatic factors than to changing subsistence patterns. Between 17 and 9 ka, increasing temperatures led to an expansion of forest vegetation. The dominant cause of fire is argued to be a weak and unstable southern-hemisphere summer monsoon and its contribution to increased frost and drought, at least in the highlands. Within the past 5 ka climate variability associated with El Niño (Sandweiss *et al.*, 1996; Rodbell *et al.*, 1999) is implicated. Haberle *et al.* emphasize that changes in fire patterns are influenced more by climatic instability than by simply dryness *per se*.

If we focus on the more recent end of this timescale, a further example of forms of agriculture which actually promote some forest

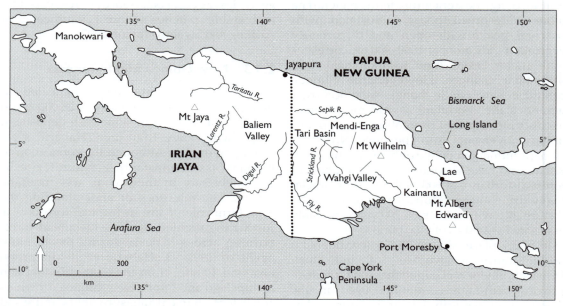

FIGURE 2.5 New Guinea location map
Source Haberle (1998: 1). Reprinted by permission of *Australian Archaeology*

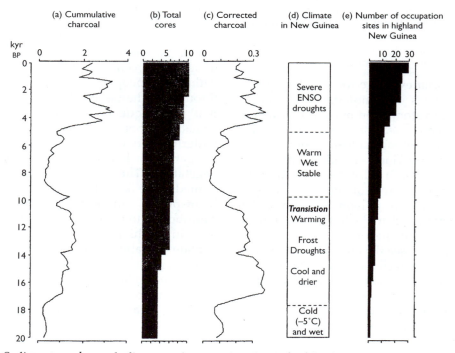

FIGURE 2.6 Sedimentary charcoal, climate and occupation sites in highland New Guinea since 20000 BP
Source: Haberle *et al.* (in press) By permission of Elsevier Science

elements comes from highland New Guinea (Haberle, 1998a), although in this example it was chosen for detailed study precisely because of its good palaeoecological visibility. This area is better known for its very early dates for agriculture, with organic infill from a drainage ditch in the Wahgi valley dated to around 10,000 BP (Haberle, 1998a: 4), although, as Haberle argues, there is no evidence for forest clearance in Wahgi pollen diagrams until about 4000 years later. (There is, however, evidence from other parts of the highlands for human impacts in the late Pleistocene (Haberle *et al.*, 1991, and *see above*).) The past 2000 years was an important period in which agricultural and pig husbandry activities and exchange networks took on their contemporary configurations. Haberle focuses on one element, *Casuarina* agroforestry, in order 'to examine its origin in space and time and to develop hypotheses regarding possible causes' (1998a: 1).

Many traces of agricultural staples, particularly sweet potato (*Ipomoea batatas*) and taro (*Colocasia esculenta*), have proved frustratingly invisible to researchers in the region. This is due to a combination of poor pollen production and preservation and vegetative means of reproduction. Haberle thus turned to proxy indicators of agricultural activity, among them the *Casuarina* pollen record. Although *C. oligodon*, which is planted for its nitrogen-fixing properties that enhance fallow regeneration and soil fertility, is not palynologically distinguishable from other *Casuarina* species, increases in *Casuarina* pollen above low background levels are 'considered to indicate development of agroforestry practices associated with firewood, fencing and soil fertility enhancement' (Haberle, 1998a: 2). In a compilation of data from 16 archaeological and 23 palaeoecological sites across the highlands, other indicators were increased forest disturbance (reduction in forest pollen types and expansion of secondary forest and non-forest taxa), increased soil erosion (increased input of inorganic sediments) and archaeological features and artefacts.

Haberle identifies at least two periods of widespread but not necessarily synchronous change within the agricultural system in the past 2000 years (Figure 2.7). *Casuarina* agroforestry first appeared around 1190–970 cal yr BP in the Wahgi and Baliem valleys, up to 400 years later in the Kainantu valley to the east,

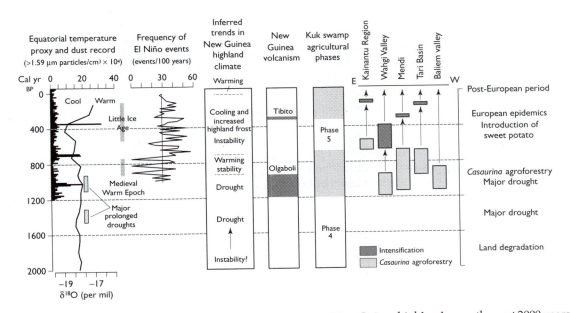

FIGURE 2.7 Comparison of climatic trends and human impacts in New Guinea highlands over the past 2000 years
Source: Haberle (1998a: 9). By permission of *Australian Archaeology*

and not until modern times in other areas. In the Wahgi the *Casuarina* increase post-dates a period of forest clearance and slope erosion. There was a second period of intensified activity after 650–400 cal yr BP, first in the Wahgi Valley. Increases in *Casuarina* planting were part of this, but raised-bed cultivation, decreased forest cover and increased soil erosion are also involved. Haberle argues this is probably the period when sweet potato was introduced to the highlands.

Haberle's comparison of climate proxies and volcanic events with the agricultural changes cautions against using temporal correlations to invoke or refute causation in a simplistic way. Nevertheless, he considers drought and volcanism to be implicated in the shift from wetland (Kuk phase 4) to dryland (*Casuarina* agroforestry) agriculture after 1200 cal yr BP (*see also* Haberle, 1998b). The relative roles of climate and other factors in the intensification phase after 400 cal yr BP are less clear-cut.

Although improved dating precision is useful, together with increased spatial resolution, the sorts of studies referred to here indicate that the most constructive use of increased precision will not be to demonstrate simple causal links between processes or events. Rather, it offers the scope to identify significant but short-lived changes, thresholds and variability that have until recently been invisible in the palaeoecological evidence. For example, Leyden *et al.* (1998) showed that, while Mayan agriculture on the Yucatan Peninsula was devastated by variable precipitation, native vegetation in the area was scarcely affected. Also important here is that such processes operate at timescales perceptible to humans; responses are likely to be the result not of selection pressure over time, but of deliberate cultural strategies.

Of particular importance in the late Holocene is climatic variability, which continues to be influential through El Niño-related droughts. Indeed, Haberle (1998a) shows cross-Pacific parallels between New Guinea and Peru in the use of dryland agroforestry, which 'may have been adopted as a response to low crop productivity and the need to rehabilitate abandoned dryland crop lands

after prolonged climatic stress' (Haberle, 1998a: 6). Understanding such interactions will be important for policy development contributing to future sustainable food production in the highlands, since 'rapidly expanding populations, increased deforestation and alienation of productive land for town and plantation use are the hallmark of present-day highlands development' (Haberle, 1998b: 10).

2.3.4 Australia

Although the Australian pollen and charcoal data set is biased in its coverage to the better-watered edges of the continent, 58 records were considered by Kershaw *et al.* (in press) to have sufficient resolution and chronological control to enable comparisons across the Holocene (Figure 2.8). Studies from a variety of environments show fire to be a consistent part of the ecosystem since 11 ka. The greatest increases in charcoal are associated with the arrival of Europeans about 200 years ago. 'This period was followed by a reduction in burning to present day levels which are, on average, lower than at any time during the Holocene' (p. 14). Within the prehistoric period there is a slight increase in burning in the past 5 ka; this is of interest as it is a period of both significant ENSO influence, as discussed above, and suggested intensification of Aboriginal occupation (Lourandos, 1983, 1993; Head, 1989).

The spatial and temporal resolution of both the palaeoecological and archaeological records is much improved since these issues were first raised in Australian debates some 20 to 30 years ago. However, all the examples discussed in this section show that we are somewhat misguided in our continuing attempts to 'decouple' or 'disentangle' human and natural causes. Instead we can productively focus on understanding in a more detailed way the environmental contexts in which humans have operated over much longer timescales.

2.4 Post-industrial impacts

Human impacts in the past few hundred years increased in variety and intensity and are

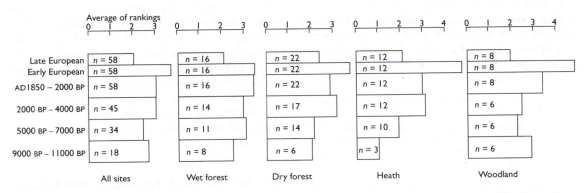

FIGURE 2.8 Relative importance of burning in time slices from the Holocene deduced from individual site rankings of charcoal abundance for 58 site records, and separately for different ecosystems, in south-eastern Australia
Source: Kershaw *et al.* (in press). By permission of Peter Kershaw

clearly evident in a range of indicators. Pollen evidence which is equivocal over prehistoric impacts is much less so over European colonization, for example in North America (McAndrews, 1988) and Australia. Temporal overviews of the more recent end of the timescale focus on the past three hundred years or so (Turner *et al.*, 1990). Within palaeoecological research, better preservation of recent evidence, larger numbers of young deposits and fine-resolution studies have facilitated a shift from inductive to deductive approaches (Oldfield, 1994). For much of the Quaternary, and for the debates discussed above, there has been little alternative to inductive approaches which aim to fill gaps in knowledge. While these may eventually lead to bodies of data amenable to the testing of specific hypotheses, that is already possible in some large-scale palaeolimnological projects. The human transformations referred to here extend, then, to climate, for example research into acid rain. This work also has the role of what Oldfield calls 'projective' research, in which the long-term perspective is used to project future change, such as under enhanced greenhouse warming (e.g. Punning *et al.*, 1997b).

As Oldfield (1994: 17) points out, deductive approaches build on an immense body of pre-existing work. In the case of the palaeolimnological studies of lake acidification, that includes detailed taxonomy and understanding of pH, salinity and nutrient preferences of organisms such as diatoms, cladocera and chironomids. It also requires dating and sedimentation processes to be well understood. Hypothesis evaluation can then often be undertaken using careful site selection, analysis of multiple proxies and appropriate statistical techniques (Battarbee, 1991).

For example, Battarbee (1991: 160–5) showed how the following competing hypotheses for lake acidification in Europe and North America could be evaluated:

(1) Lakes may be naturally acidic and have not changed over time.
(2) There has been slow post-glacial acidification as a result of leaching.
(3) There has been acidification since 1800 as a result of land-use changes including increase of conifer forest and decline of burning and grazing.
(4) There has been acidification since 1800 as a result of increased acid deposition from fossil fuel combustion.

The first two hypotheses were relatively easily ruled out, since 'evidence for a marked reduction in lake-water pH is usually confined to the most recent sediments, postdating 1800 a.c.e. [after the Christian era, i.e. AD 1800] and usually following long periods of very stable conditions' (Battarbee, 1991: 161). A range of studies ruled out hypothesis 3. Adjacent nonafforested and afforested sites show acidification over similar timescales, or that acidification preceded planting. Separating hypotheses 3 and 4 is more difficult when the timescales of land-use and acid deposition changes are the same.

> However, it is possible to remove the influence of acid deposition either by examining a sensitive site with land-use change in a 'clean' area or by examining the lake response to an analogous change in soil or vegetation in the 'clean' (pre-1800 a.c.e.) past. Such a 'past analog' can be identified by pollen analysis ...
> These data show that land-use changes in the absence of acid deposition have little effect on lake acidity; however, land-use changes in association with acid deposition can be important. (Battarbee, 1991: 161)

I have oversimplified this example to contrast the explanatory power of this type of study with the more inductive work on earlier periods, discussed earlier in the chapter. As the hypothesis becomes more precisely defined, it may allude to only part of a wider set of environmental issues. For example, while conifer afforestation in the UK is not implicated in lake acidification, it does have other detrimental effects, such as accelerated soil erosion through intensified catchment interference, and nutrient enrichment associated with fertilizer and pesticide use (Battarbee *et al.*, 1985; Battarbee, 1990). Of course, high-resolution work in the more recent period also has the potential to show much more complex patterns of interaction and feedback. Industrial activities influence atmospheric, biological and geological processes in increasingly complex ways; simple relationships between cause and consequence should not be expected (e.g. Punning, 1994). A further feature of deductive approaches is the necessity of large research teams combining various skills and techniques.

Impacts assessed by a team from the Institute of Ecology in Estonia include those associated with the largest commercially exploited oil shale deposit in the world. Excavation of the oil shale leads to groundwater depletion and deforestation, and shale oil production results in atmospheric emission of fly ash (Punning, 1994). As well as indicators such as pollen, diatoms and charcoal, trace elements, spher-

oidal fly-ash particles and heavy metals have been examined (Varvas and Punning, 1993; Punning *et al.*, 1997a). Historical data also becomes an important research tool in this context (Koff *et al.*, 1998). The integration of multi-proxy palaeoecological indicators with historical evidence is a feature of studies of recent human transformations in diverse corners of the world (e.g. Baker *et al.*, 1993b, in the midwestern USA).

2.5 Strengths, limitations and future directions

There are definite trends perceptible in the diverse examples included in this chapter. In all debates there is increasingly explicit attention to working at spatial and temporal scales appropriate to the questions being asked. In the case of questions about human impacts, local-scale studies are of particular importance. At the recent end of the time spectrum this work has led towards explicit testing of hypotheses. However, even in earlier periods in which not all variables can be controlled for, comparisons between different scales are producing important insights. In turn, more nuanced explanations can be developed, even if they are still within an inductive framework.

With the influence of a simplistic environmental determinism virtually completely removed, it has become possible to discuss the agency of the environment in more complex and interesting ways. Simple correlations with environmental variables, whether to invoke or to rule out causation, are spurious, and there is increasing interest in thresholds, perceptions and variability.

Some workers have argued that our thinking has been unduly constrained by the nature of the evidence to hand. For example, dominance of pollen and charcoal evidence in these debates has meant a focus on vegetation. While many would still argue that fire was part of the earliest human toolkits, it is important to think about the likely impacts of activities for which we have no evidence. The very clear trend to the use of multiple proxies will continue. Further, once we are no longer attempting to deduce multiple processes from single or few indicators (for example, climate, ecological process and human impacts from pollen), each indicator can be used more effectively and appropriately.

Similarly, interpretations have been heavily biased by the regions where most research has been undertaken. With complex human histories and still limited palaeoecological work, South-East Asia and Africa stand out as regions that are likely to reorient our thinking in the next few decades.

As it is, the studies discussed do not just offer an increasingly detailed understanding of the ways in which humans have transformed the Earth over the long term. They also provide challenges to a number of debates within the social sciences. For example, very dichotomized views of hunter-gatherer and agricultural systems, and deterministic trajectories of evolutionary change from one to the other, are not supported by the evidence discussed here. The move away from very large scales of analysis that conflated a great deal of social and ecological complexity facilitates more systematic thinking about similarities and differences. Not just across the Australia–New Guinea divide, but more widely, we seem to be moving away from the assumption that hunter-gatherers automatically have certain sorts of (little) impacts and agriculturalists have others (big). There are connections between the palaeoecological perspective on these questions and the archaeological thinking (Chapter 4).

The apparently messy colonization stories, such as the New Zealand one, have the potential to subvert very linear notions of human history, of conquering the corners of the Earth. Even though it might be resolved in a non-messy way, it has great promise for a more complexly rendered understanding of the human history of the Pacific. What happened after deforestation? What are the permutations of land use? What were the relationships between social structures and land use? In helping us to think about colonization as a process rather than a single event, the record of human transformations also connects with conversations occurring in the social sciences, as will be discussed in Chapter 5.

The question of naturalness: environmental change in ecology and palaeoecology

3.0 Chapter summary

Ecology and palaeoecology have undergone something of a convergence in recent decades in a renewed emphasis on historical rather than cyclic change. The concept of the balance of nature has been found severely wanting. Approaches referred to as 'non-equilibrium ecology' draw on empirical evidence of change, instability and dynamism in ecosystems, and challenge Clementsian ideas of succession, stasis and equilibrium. Palaeoecological evidence of long-term vegetation change, species- rather than community-level response and variable refugial strategies supports the same conclusions from different directions. In both fields, disturbance is now understood as integral to ecosystems rather than an external process. Change is the norm. In management terms, that does not mean that 'anything goes', as we are now in a better position to understand the frequency, magnitude and degree of change from different causes. This work raises both philosophical and practical management challenges to concepts of naturalness, pristine baselines and human impacts. It also provides exciting practical opportunities for the past to inform the present and future, since the past influences contemporary ecosystems at a range of timescales. These opportunities are more likely to be at the local and regional than the 'big picture' scale.

3.1 The change in environmental change

This chapter is about the 'change' in environmental change. Its central theme is the shift in understanding – from stability as the norm to change as the norm – that characterizes the ecological sciences in recent decades. This rethinking is often referred to as 'the new ecology', or 'non-equilibrium' or 'disequilibrium' ecology. The long-term record of changing environments is relevant in the first instance irrespective of what it says about humans. Research is consistently showing that 'change takes place all the time, in all sorts of directions and at all sorts of scales, catastrophically, gradually, and unpredictably' (Stott, 1998: 1). Integral to this rethinking is a shift from the cyclical time of systems ecology to historical time (Zimmerer, 1994), and thus the exciting possibilities of linking the deep past to contemporary questions which are the subject of this book. The 'new ecology' requires us to critically rethink concepts important to our broader discussions. What is natural? What is the normal state of affairs in the environment? Where are the baselines, and is the concept of a baseline even relevant? Inevitably, then, the view that change is the norm has human implications: first, for a different perspective on the human role in environmental change, and second, for management of environments and landscapes. Stott argues that many of our management strategies are directed to the impossible, indeed spurious, goal of stability. 'We are trying to replace human flexibility and adaptation by god-like control and stasis, and it will not work' (Stott, 1998: 1).

Work in the natural sciences provides a very clear picture of changing environments. It is possible to build up a relatively precise picture of the multiple spatial and temporal scales at which change occurs (Figure 3.1). Concepts such as periodicity, discontinuity, trends and rates are integral to these discussions (Huggett, 1997). Awareness of long-term environmental change is not a new phenomenon, but was central to the nineteenth-century revolutions in geology and biology. However, the dominance of systems models within ecology throughout the middle decades of the twentieth-century led to ahistorical approaches and an emphasis on notions of stability and equilibria.

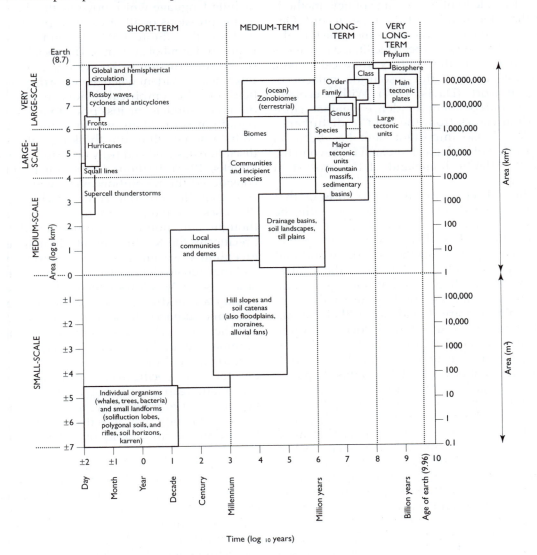

Figure 3.1 Spatial and temporal scales of environmental change
Source: Huggett (1997: figure. 1.2, p. 5). By permission of Taylor & Francis Books Ltd

The evidence in this chapter is drawn from two subdisciplines that have traditionally operated over different timescales and within different frameworks. They are now, however, showing a clear convergence on these issues. Quaternary palaeoecology operates mostly at a thousand-year timescale, while ecology's main framework is decadal. What is new in thinking about deep time is our ability to assign precise dates to particular changes, using a range of radiometric and other dating techniques. We are also able to utilize a range of new methods in the interpretation of the fossil record. One aspect of the convergence, a more historically oriented ecology, has been facilitated by a focus on change at the century timescale made possible by new fine-resolution palaeoecological work (Davis, 1994). Within ecology itself, independent evidence against equilibrium comes from the field of disturbance ecology. Also important here has been an expansion of ecological research far beyond the North Atlantic homeland of the ecosystem concept.

These discussions take us to an issue that many scientists are uncomfortable with: the way in which the social context of the production of knowledge enhances the power of some depictions of the material world and disempowers others. Although the central theme of this chapter is that non-equilibrium views have gained ground because of the overwhelming empirical evidence in their favour, this has not occurred in a social vacuum. Further, we need to ask why it has taken so long for change to be accepted as central. Perhaps what is most notable in this chapter is the fact that in the late 1990s the authors of keynote papers, such as Stott (1998) and Meadows (1999), have felt such a strong need to reiterate the message of change. This, more than anything else, reminds us how entrenched the metaphors of balance, harmony and stability have become. The chapter, then, requires us critically to consider the ways in which ecology and palaeoecology construct stories about the world, ideas which are also important in Chapters 4 and 5. In the final sections of the chapter we consider the implications of non-equilibrium views for human

questions. These are issues which will be taken up more specifically in the second half of the book.

3.2 Ecology

It perhaps surprised some scientists when the editor of the *Journal of Biogeography* recently attributed the crisis in biogeography and ecology to issues of metalanguage – 'the overarching language which governs the thoughts and expressions of all of us, both professional and popular' (Stott, 1998: 1) – and semiotics (signs and symbol systems). The social context of the construction of knowledge is perhaps most clearly evident when there is a disjunction between the empirical evidence and the dominant models, or just prior to a Kuhnian paradigm shift. Many would argue that the shift to non-equilibrium thinking within the discipline has well and truly occurred (Wu and Loucks, 1995). Barbour (1996), for example, considered the revolution over by 1960. It is true that the broader context which Stott discusses is at the interface between academic thought and public environmental debate, specifically the Kyoto Summit discussions on greenhouse warming and the possibility of curbing climate change. However, his writings and those of Meadows (1999) (*see* discussion below) are addressed very explicitly to an academic audience. In Stott's analysis the 'key signifiers' – climaxes, optima, balance, harmony, equilibria, stability, ecosystems, synecology, and the 'exotic other' – have been developed in specific historical circumstances. The main features of that context are the hegemony of forest ecology; the hegemony of equilibrium notions; and the hegemony of Europe and North America over the rest (Stott, 1998: 1).

This chapter cannot do justice to the long and complex history of ecological thought, and any summary risks oversimplifying the issues. Hagen (1992: 11) has emphasized that the dualistic themes under discussion here – 'change and uniformity, instability and equilibrium, competition and cooperation, integration and individuality' – were established by about

1900. In many ways non-equilibrium ecology can be thought of as the most recent expression of debates that go back to Frederic Clements and Henry Gleason in the early decades of the twentieth century. At issue then were the status of the vegetation community, and the ideas of succession towards climax.

> Clements argued that groups of species living together in a given habitat were highly organized into natural, integrated units called communities. Gleason countered that such communities were only constructs of human thought and that in reality the distribution and behavior of every species were unbounded by imagined holistic bonds to all the surrounding species. (Barbour, 1996: 234)

Despite a long-standing critique of succession, it has been extremely powerful as an image and as an idea (Figure 3.2).

There were also important regional contexts to these debates, and their links to conservation (Bocking, 1997). For example, ecologists such as Arthur Tansley, who coined the term 'ecosystem', were influential in the development of nature conservation in Britain. They argued that comparison of 'manipulated' and 'natural' (i.e. reserved) ecosystems was important for advancing ecological knowledge (Adams, 1997). In the context of urban expansion, this nature was seen as 'essentially static – an array of habitat fragments as natural objects set in a landscape of change' (Adams, 1997: 281).

In the USA systems ecology was particularly associated with the work of Howard and Eugene Odum. Also influential were watershed-scale studies such as the Hubbard Brook study begun by Bormann and Likens (Hagen, 1992; Bocking, 1997). On both sides of the Atlantic there was an interest in applying the results of ecological research to management and conservation questions (Bocking, 1997). Management was conceptualized in interventionist terms (Adams, 1997: 283), reinforcing an understanding of humans as external to the system being studied. There is also a regional influence on the recognition of non-equilibrium ecology, the dynamic nature of change being particularly evident to workers in arid lands (e.g. Sullivan, 1996) and developing countries (Zimmerer and Young, 1998).

Empirical challenges to homeostasis and equilibrium emerged particularly in the 1970s, mainly through recognition of the dynamism

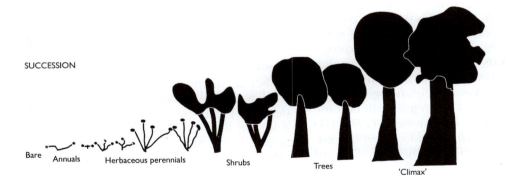

SUCCESSION

Bare   Annuals   Herbaceous perennials   Shrubs   Trees   'Climax'

FIGURE 3.2 The Clementsian model of succession after disturbance in natural landscapes, as reproduced and critiqued by Blumler (1998: 222). Note particularly Blumler's caption: 'Similar, but more beautiful, diagrams can be found in all ecology texts. Such diagrams idealize and therefore encourage acceptance of the model, whereas the sketch here trivializes it, and therefore makes it easier to perceive the underlying assumptions of progress, linearity, equilibrium, and bioutopia (climax)'
Source: Karl S. Zimmerer and Kenneth R. Young (eds), *Nature's geography*. © 1998. Reprinted by permission of The University of Wisconsin Press

and non-linearity seen in disturbance ecology (Pickett and White, 1985a). Two themes are particularly evident in influential writings on the so-called new ecology. The first is that new ways of thinking and talking were advocated by workers on the basis of empirical discrepancies: what they saw did not match the models. For example:

> The job needs doing because I am uneasy with the repeated assertions that nature's norm is balance, that this balance is fragile, and that current human activities invite the collapse of entire, complex ecosystems. In contrast, what I have seen during decades of fieldwork is neither pervasive order nor chaos, but comfortable disorder (Drury, 1998: 1).

Second, it was clear to Drury and others that there were profound implications here for conservation and management. There is a strongly conservationist line in this literature, but the implications are not self-evident and require rethinking of static views of conservation. These issues are discussed most fully by Botkin (1990), who challenges the idea of nature as stable, balanced and undisturbed. As Botkin (1990: 10) asks, 'How do you manage something that is always changing?'

Critics point to two main problems in accepting the new approaches (*see* Zimmerer, 1994 for review). First, it is argued, placing pluralism and contingency at centre stage undermines the possibility of ecology providing unified explanations. Second, research will become smaller scale in terms of the problems it takes on, because generalizations about complex systems will become much more difficult. As Zimmerer (1994: 111) argues, these are difficulties to be addressed, rather than strong attacks on the truth claims of the new ecology.

3.3 Palaeoecology

3.3.1 Species and community

Barbour (1996) attributed the shift away from the Clementsian and towards the Gleasonian view of the plant community to complex cultural forces by which knowledge was produced in North American plant ecology.

This 'revolution' of the 1940s and 1950s, he argued, was the product not of an accumulated weight of new evidence but of cultural factors conducive to the idea of fragmentation. Drawing on the oral histories of practitioners, his survey

> identifies widespread American cultural themes of fragmentation [*sic*] of norms, the celebration of the individual, and rebellion against convention. Impermanence and uncertainty replaced predictability, and individual competition displaced group cooperation. At the same time there was an expression of anxiety at the loss of past certitude and stable social organization. (Barbour, 1996: 250)

Without downplaying the influence of broader cultural influences on scientific thought, it is interesting to note how comprehensively the actual evidence from palaeoecology (mainly pollen) has demolished essentialist ideas of vegetation communities in the decades since 1960. This is particularly the case in North America, where pollen analysts have been able to map species movements following the last glacial retreat in considerable detail (*see*, for example, Davis, 1983; Delcourt and Delcourt, 1987) (Figure 3.3). Three factors have facilitated this mapping: first, the wide distribution of sites with conducive preservation conditions; second, the detailed work generated by a large research population; and third, the fact that northern-hemisphere forest formations are represented palynologically by a useful number of distinctive tree genera. In more recent work the pollen mapping is corroborated and given greater spatial and temporal resolution with the use of plant macrofossils (Jackson *et al.*, 1997).

The individuality of response of plant species to environmental perturbation is indicated in much recent work in other parts of the world (see Meadows, 1999 for review). In the Amazon, altitude-determined vegetation belts did not respond uniformly to glacial and postglacial conditions (Colinvaux *et al.*, 1997). 'Rather, plant associations formed and reformed as species occupied new centres of distribution in response to fluctuating climates' (Meadows, 1999: 259). Similarly, in highland central Africa, time-transgressive patterns of

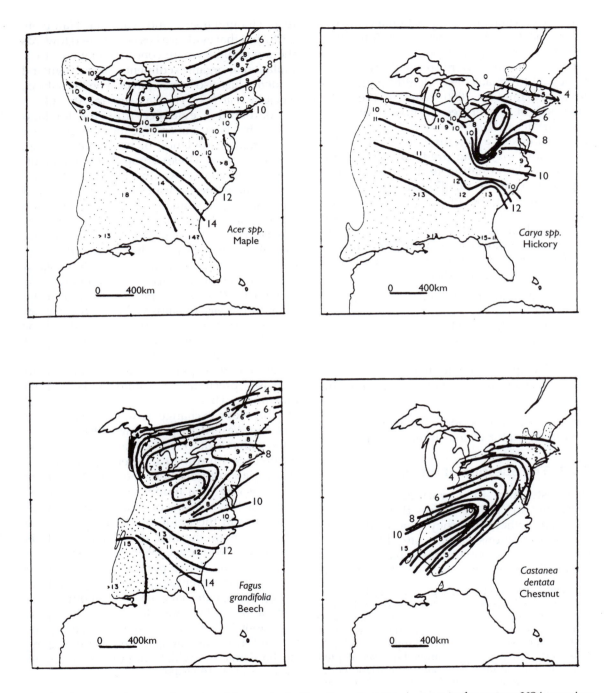

FIGURE 3.3 Maps showing the times of first arrival of four boreal species/genera in the eastern USA, moving northward during the late Wisconsin and early Holocene. Small numbers indicate arrival times in thousands of years at individual sites; isopleths connect points of similar age and represent the frontier for the species at 1000-year intervals. Stippled areas are modern ranges

Source: Davis (1983: figure 11.7, p. 175). By permission of The University of Minnesota Press

forest expansion and compositional changes in Holocene forests are indicated (Jolly *et al.*, 1997). There are exceptions, however. In African lowland rainforests further west, Maley and Brenac (1998) identify periodicities of approximately 2000 years in the pollen curves of evergreen taxa such as Flacourtiaceae, *Diospyros* and *Trichoscypha*. In that context 'a modified "Clementsian" succession, involving predictable vegetation sequences culminating in a climax, remains arguably appropriate' (Meadows, 1999: 60). Further, in a number of examples reviewed by Meadows, plant communities identified palaeoecologically have no modern analogue.

A single 'rainforest' is at any one time an artefact of biotic response to multiple past disturbances of different periodicities, for example:

- 10–100 years for regeneration of tree fall gaps;
- 100 to at least 500 years for response to favourable climate change;
- 10–100 years for setback due to unfavourable climate change;
- 1–10 years for recovery from large fires (Maley and Brenac, 1998).

Another series of periodicities would need to be considered for various human influences.

3.3.2 The question of refugia

If forests and their various components expanded in complex ways throughout the Holocene, where had they been in the harsh glacial conditions of the late Pleistocene? Debates over potential refugia have long and regionally variable histories. For the tropics, a

> viewpoint arose that interpreted areas ('islands') of elevated diversity and levels of endemism within the tropics as the Pleistocene refugia of rainforest taxa. According to this view, these areas represented places of refuge at moderate altitude, where plant and animal species adapted to the hot and humid climates of lowland equatorial regions survived during what were emerging as cooler and drier climates of the glacials, when the savannas became much more widespread. (Meadows, 1999: 258)

Meadows points out that this view has been criticized on the basis of palaeoecological evidence

(e.g. Hooghiemstra and Van der Hammen, 1998), but that evidence has been biased towards the mountains, where conditions favour accumulation of organic sediments. (Some highland evidence has also been equivocal on the subject, e.g. Jolly *et al.* (1997) for East Central Africa.)

Comparative evidence from the lowland tropics has begun to emerge more recently. For the Amazon lowlands of north-west Brazil, Colinvaux *et al.* (1996) argue that even significant glacial cooling was insufficient to displace or fragment the rainforest significantly, and alternative explanations for species endemism are indicated. For lowland Africa 'the refugium theory appears still to offer an appropriate model of vegetation change in the late Quaternary, even if it does not offer a valid explanation of patterns of diversity and endemism' (Meadows, 1999: 260). And, if the refugium theory remains viable for the African forests, it is partly because of some evidence for glacial forest expansion. DeBusk (1998) argued that montane forests around Lake Malawi expanded in response to both cooling and increased precipitation around the last glacial maximum. This supports a 'stepping-stone' view of biotic exchange between the forests of East and West Africa (Meadows, 1999: 262).

On either side of the equatorial rainforest biome in South America, vegetation instability and variation throughout the late Quaternary are indicated by a number of recent studies. Thus 'the key message of Quaternary biogeography lies in the ubiquity of change and the fallacy of assuming that, in nature, stability is normality' (Meadows, 1999: 266). In terms of understanding the response of ecosystems to future environmental, particularly climate, change, the Quaternary record has both limitations and a vital role (Box 3.1). While past conditions in themselves offer poor analogues for the future, the key is in understanding the mechanisms by which organisms respond (Huntley *et al.*, 1997: 489).

3.4 People and disturbance

The rise of non-equilibrium approaches both requires and facilitates a rethinking of the way

> **Box 3.1 The Quaternary fossil record as a key to the future: limitations and strengths summarized from Huntley et al., 1997)**
>
> **Strengths**
>
> - Longer timescale than contemporary studies.
> - Identify potential refugial areas.
> - Provides data to test model predictions.
> - Future responses of organisms will be by the same principal mechanisms, and be subject to limitations inherent in these mechanisms.
>
> - Prediction of animal taxa at greatest risk of extinction.
>
> **Limitations**
>
> - Late twentieth-century human impacts are without precedent.
> - Trace-gas-induced future climate change may differ in pattern and magnitude from orbitally induced change.
> - Altered atmospheric composition (especially CO_2) will have impacts on plant function.
> - Taphonomic and taxonomic biases in fossil record.
> - Variable spatial and temporal resolution.
> - Human impacts have altered potential migration areas.

people are viewed within ecology (McDonnell and Pickett, 1993a). This is illustrated here by a comparison with 'disturbance'; metaphorically, people and disturbance are similarly placed in this debate. Are they normal and part of the system, or external? For example, Adams (1997: 286) argues that 'gone … are the days when conservationists could conceive of "nature" in equilibrium and hence portray human-induced changes in those ecosystems as somehow "unnatural".' We could replace 'human' in this quotation with 'disturbance' or one of its variants, 'fire', and make exactly the same argument. Perhaps surprisingly, the ecological literature continues to use the term 'disturbance', with its connotation of interference with the due course of any action or process, at the same time as it is shown to be an integral part of ecosystems (e.g. Klomp and Lunt, 1997: 53–4; Peacock *et al.*, 1997). Much global change research also uses definitions of disturbance as external to the system (e.g. Sala *et al.*, 1999: 309).

Further, as White and Pickett (1985) argue, it is not just a matter of reversing existing categories of exogenous and endogenous change. Rather, because distinctions between 'inside' and 'outside' are difficult to make, they can be regarded as the endpoints of a continuum, with many disturbances combining aspects of each.

For example, 'senescence and death of over-story trees is an intrinsic community rhythm, but the proximal cause is often windstorm' (White and Pickett, 1985: 8). Other important features of disturbance have parallels in human activity: it is often patchy; it is distributed in time (frequency, discreteness, rotation time); it varies in impact or magnitude; and two agents may act synergistically, as well as interact with stress (Pickett and White, 1985b).

This raises the whole issue of managing for change, and the dilemmas between intervention and preservation (Drury, 1998: 187–92). 'Natural disturbance can be seen as part of the "nature" with which conservation is concerned' (Adams, 1997: 286). The issue of change is to the fore in contemporary management of forests, rivers, coastal environments and storm-damaged woodlands. Some of these are examined in more detail in Chapter 8.

There are profound and difficult questions here in terms of management. 'Removing the anthropogenic factors of change does not return us to a "stable" environment, for the earth's environmental history is one of continual variability and adjustment' (Meadows, 1999: 266). Adams (1997: 286) suggests that 'the environmental manager has now perhaps to be more croupier than engineer'. The predictive capacity of the non-equilibrium approaches is weaker (Zimmerer, 1994), but perhaps more realistic in being conscious of its limitations. There are also fears that acknowledgement of change may lead to an 'anything goes' approach and provide further justification for human-induced degradation. However, 'to accept certain kinds of change is not to accept all kinds of change' (Botkin, 1990: 11). Well-developed senses of scale, and critical understandings of the way knowledge is produced and utilized, are important variables in constructively using the insights of the 'new ecology'. Many of the readers of this book will be in a good position to compare different types of change in terms of frequency, magnitude and degree (Zimmerer, 1994; Zimmerer and Young, 1998).

There is also a key paradox. The dominance of humans shows how integral we are to environmental change, yet our externality is reinforced by the conceptualization of human influences in terms of 'transformations of the Earth' or 'alteration of Earth'. For example, in arguing that 'even on the grandest scale, most aspects of the structure and functioning of Earth's ecosystems cannot be understood without accounting for the strong, often dominant influence of humanity', Vitousek *et al.* (1997: 494) have to conceptualize humanity as partly integrated with, but mainly external to, the Earth system (Figure 3.4). The finding of human dominance is based on a review of a large body of well-known empirical evidence which is used to summarize the human contribution to environmental change in quantitative terms (Figure 3.5). Issues raised by this paradox are explored further in Chapter 5.

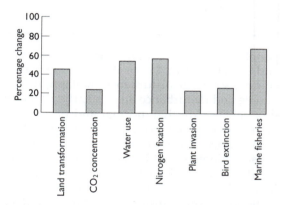

FIGURE 3.5 Human dominance or alteration of several major components of the Earth system, expressed as (from left to right) percentage of the land surface transformed; percentage of the current atmospheric CO_2 concentration that results from human action; percentage of accessible surface fresh water used; percentage of terrestrial nitrogen fixation that is human-caused; percentage of plant species in Canada that humanity has introduced from elsewhere; percentage of bird species on Earth that have become extinct in the past two millennia, almost all of them as a consequence of human activity; and percentage of major marine fisheries that are fully exploited, overexploited, or depleted
Source: Reprinted from P.M. Vitousek, H.A. Mooney, J. Lubchenco and K.M. Melillo, Human domination of Earth's ecosystems. *Science* **277**: 494-9 (figure 2, p. 495). Copyright 1997 American Association for the Advancement of Science

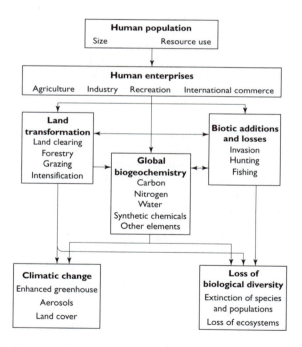

Figure 3.4 A conceptual model illustrating humanity's direct and indirect effects on the Earth system
Source: Reprinted from P.M. Vitousek, H.A. Mooney, J. Lubchenco and K.M. Melillo, Human domination of Earth's ecosystems. *Science* **277**: 494-9 (figure 1, p. 494). Copyright 1997 American Association for the Advancement of Science

Not all human impacts are dramatic and obvious. Indeed, the more people are integrated into ecosystems, the more complex are the processes likely to be. Thus the conceptualization of 'subtle human effects' is particularly important (Table 3.1). The examples given in the table tend to be negative ones, but there is no reason why subtle human effects are necessarily destructive.

The examples in the table show very clearly the temporal dimension of subtle effects. They are often subtle precisely because the causal mechanism or trigger happened some time ago. Although some can be global, the spatial and temporal scales at which many subtle impacts operate is comparable to the analytical scales of ecological, palaeoecological and archaeological research, as we saw in some of the examples of sections 3.2 and 3.3. It may be that it is in this domain that understanding of long-term environmental change will be of most practical management use.

3.5 Change in global environmental change research

The burgeoning field of global change research is predicated on the ubiquity and acceleration of change. While it builds on work in the two disciplines already discussed, as well as others, it is important to consider how change is conceptualized in this specific area. The International Geosphere–Biosphere Programme (IGBP) identifies three main components of global change: land-use/cover

Table 3.1 Examples of subtle human influences within ecosystems

Effect	Definition	Example
Indirect	Outcome of interaction between two entities is mediated by third party	Introduction of organism that changes trophic structure
Historical	Alteration of a contemporary interaction resulting from prior state	Forest structure as a result of land use a century ago
Biological	(subset of historical) Persistent alterations in biotic component	Deposition of organic layer in soil by partial vegetation harvest
Echo of the past	Historical effect only sporadically apparent	Destruction of a species refuge used only during drought
Lagged	Triggered some time before they appear	Gradual destruction of stratospheric ozone by chlorofluorocarbons
Unexpected action at a distance	Can be unsubtle, e.g. pollution release, but more unexpected when they are subtle	Impact of pest eradication programmes on migratory animals

Source: McDonnell and Pickett (1993b)

change; changes in atmospheric composition; and climate change. Humans are seen as a fundamental driving force in each of these, and the dominant influence on land-use/cover change. In a fourth area, changes in biological diversity (considered as both a component and consequence of global change), humans are also a key force (Walker and Steffen, 1999). As Table 3.2 shows, the influence of people is predicted to exceed that of climate change on terrestrial ecosystems in all biomes except the tundra–boreal forest.

The temporal dimensions of humans and disturbance are clearly recognized in this research, since it has shown very clearly that most of the world's vegetation is not in equilibrium with present climate. Thus 'past disturbance will play an important role in determining the future composition of long-lived plant communities' (Walker *et al.*, 1999: 368).

The recognition of change is now such that some critics have labelled 'sustainability' and 'equilibrium' as 'dangerous concepts' in the global environmental change debate:

Table 3.2 Predicted dominant driving force and critical management issues in four latitudinal zones over the next century

Zone (biome)	Predicted dominant driving force	Critical issues
High latitudes (tundra–boreal forest) (Alaska, USA/Canada, Siberia)	Climate change – significant temperature increases in past three decades	Sustainable management of boreal forest – logging methods need reappraisal for regrowth in warmer drier conditions
Mid-latitudes (grassland and deciduous temperate forest) (North America, China)	Land-use change	Land use and agricultural sustainability. Climate change may increase vulnerability to rare climatic events, erosion, pests (but currently unpredictable)
Sub-humid to semi-arid tropics (tropical savanna) (West Africa, Australia)	Land use likely to override CO_2 and climate change for several decades	High levels of grazing and/or cultivation increase sensitivity to changes in climatic extremes
Humid tropics (tropical rainforest) (Amazon, Sumatra)	Land use	Pressure for conversion of forest to agriculture, and for commercial forest harvesting will increase through twenty-first century

Source: Walker *et al.* (1999)
Note: Regional analysis based on case studies from areas indicated in brackets.

In these concepts, the idea of change is accepted, but it is then immediately corrected and ameliorated by the qualifying notion that the direction and degree of change are 'sustainable', and not self-destructive, and that systems are likely to settle down to an original, or to a new equilibrium level, at a certain point in the future. These palliatives come in many versions (Moore *et al.*, 1996: 199).

Whether this is a fair summary of global change research is debatable, as is the meaning of 'sustainability'. These issues take us into the cultural domain of the next two chapters. I shall have more to say about how global change research conceptualizes culture, and the cultures of global change research, in Chapter 5.

Part III

METHODOLOGICAL AND CONCEPTUAL TOOLS FROM THE HUMANITIES

4

The social construction of nature and landscape

4.0 Chapter summary

Within the humanities, recent rethinking of central concepts has taken place in a context influenced by the diverse perspectives of postmodernism. Foundational narratives, including those of science itself, have been under challenge. The challenge needs to be met rather than ignored or dismissed. Rigorous attention to the conditions under which knowledge is produced, and to the cultural or historical context of influential explanations, can only lead to improved understanding. The recent social constructions of culture, nature and landscape are explored. All are shown to have multiple and shifting meanings. They converge with evidence presented in Chapters 2 and 3 in demonstrating the problematic status of key environmental concepts such as wilderness and pristine nature. The use of the landscape concept in prehistoric archaeology is discussed in relation to debates about the European Neolithic. This provides a long-term perspective on several key themes: multiple meanings of landscape; the science/humanities dialectic; the material expression of symbolic behaviour; the prehistory of landscape memory; and the long history of ascribing meaning to the nature/culture dualism.

4.1 Culture, nature and landscape

In Chapter 2 we looked at how cultural landscapes have been understood, within a section of the natural sciences, as those physically transformed by human action. At least implicit, and sometimes explicit, in these analyses is the conceptual separation of culture and nature. Researchers may have trouble attributing particular evidence to either cultural or natural causes, but in general they do not regard the dualistic categories of human and nature as problematic in themselves. In Chapter 3 we gained more of a sense of the problem in debates over what to do with humans in conceptualizations of ecosystems. Within the humanities, however, central concepts such as *culture*, *nature* and *landscape* are seen not as preexisting realities but as categories constructed by social processes. Their meanings are multiple and shifting.

In this chapter we examine shifts in the understanding of these three concepts within the humanities in the past few decades. This is not to suggest that previous conceptualizations were fixed; on the contrary, they have had a long history of change even purely within Western thought (Glacken, 1967). However, there are some distinctive features of the recent context which are important to understand. We start with a brief overview of that context, then examine what it means for the concepts of culture, nature and landscape. This does not mean that they are only concepts, with no material expression. Neither does it mean that nature has no agency independent of humans. Attention to the social dimensions of landscape has been of particular concern in archaeology, and is considered in some detail.

The so-called 'cultural turn' in the humanities is often associated with postmodernism (*see also* Box 4.1), a movement 'characterized by scepticism towards the grand claims and grand theory of the modern era, and their privileged vantage point, stressing in its place an openness to a range of voices in social enquiry, artistic experimentation and political empowerment (Johnston *et al.*, 1994: 466). An emphasis on difference and diversity, rather than universality, in explanation is one of the hallmarks of the humanities in the past few decades. Awareness of that diversity comes from a range of sources. For example, feminism has been pivotal in showing how supposedly universal explanations are gendered; and non-Western and indigenous voices have challenged the ethnocentrism of Western thinking.

As Cosgrove argues (1990: 344), the term 'postmodernism' implies a closure to something thought of as modernism. Although in some ways just as broad an umbrella term as postmodernity, modernity shares the characteristic of intellectual and historical trends being intertwined. (Some of the confusion engendered by the use of the terms in relation both to attitudes and to periods of time is discussed in the box.) The Modern era is usually associated, then, with:

- the Scientific Revolution (the rise of rationality and objectivity in knowledge systems);
- predominantly European colonialism, in which exploration and the development and extension of capitalism were linked;
- 'a belief in historical linear progress through technology and materialism' (Cosgrove, 1990: 351).

These strands were linked into an understanding that the world was knowable, using rational, empirical means, an understanding that persists in much positivist science today.

The new science replaced previous analogical and metaphorical ways of knowing. It 'distrusted metaphor, or, more precisely,

---

Box 4.1   Putting postmodernism in a box

In view of its diverse meanings and usage, and its emphasis on the diversity of speaking positions, *postmodernism* is perhaps the least appropriate term in the whole book to put in a box! The following material is summarized from Johnston *et al.* (1994).

There are three usages of the term 'postmodernism' that Dear (1986) found it useful to distinguish:

- There is postmodern *style*, in which architecture is the paradigmatic art, characterized by a diversity of design elements.
- Postmodern *method* centres on the strategy of *deconstruction*, which seeks to demonstrate how the multiple positioning of the author and reader (for example, in terms of gender, class, historical background) influences the writing and reading of text. Deconstruction is a destabilizing method that challenges claims to truth and looks for ambiguities, incoherences and alternative readings.
- Postmodern *epoch* is a phrase often used in reference to the present era, characterized by rapid change and globalization. Harvey (1989) has argued that we are not actually 'post'-modern; rather, we are in the most recent stage of capitalist evolution, characterized by flexible accumulation and space–time compression.

A related movement is *post-colonialism*, which challenges the impact of imperialism on non-Western cultures. Post-colonial writing illustrates the ethnocentrism of much Western history, geography and literature, and provides alternative voices, e.g. indigenous ones. In a similar way, feminism has shown the gendered nature of much Western thinking.

regarded its metaphors and models as transparent lenses to true meaning rather than themselves constitutive of it' (Cosgrove, 1990: 350–1).

The critique of scientific claims to truth gained ground in a late twentieth-century social context of pessimism about both science and progress. Not only had science not prevented significant environmental problems, it was deeply implicated as part of the problem in issues such as nuclear weapons, acid rain and chemical pollution.

Critics – in both the sciences and the humanities – of postmodern approaches and cultural theory focus on two main issues: inaccessible language and relativism. In suggesting ways in which human geographers can engage with practising scientists in these debates, Demeritt (1996) reminds us that, with a few exceptions, the sciences and humanities have continued to occupy parallel universes. 'Ironically, bold pronouncements by post-modernists of the death of Enlightenment metanarratives have been written in such a way that the next of kin – empiricist and positivist physical geographers – missed the funeral' (Demeritt, 1996: 485). Demeritt puts the onus on social constructivists to communicate better, and to focus on the analysis of scientific practice rather than more abstract discussions of representation (*see* Chapter 5 for more detailed discussion). However, there is also an onus on scientific practitioners, and environmental scientists in particular, to engage with these debates. It is unfortunate if arcane language provides an excuse for avoidance. If the term 'postmodernism' is a barrier to the readers of this book, they are welcome not to use it, but there are issues here which cannot be avoided.

For many scientists, challenges to the possibility of objective truth are unnerving. This is particularly the case given the close links between Western scientific perspectives on conservation and global change research. Where does one find a defensible political position if all these diverse perspectives are equally valid? The most creative thinking about the dilemmas of objective truth, social constructivism, relativism and political positioning that I have found is in feminist critiques of science

(Harding, 1986; Haraway, 1991). They have important implications for what is perhaps the central dilemma of this book: how are we to harness a long-term perspective on culture, nature and change – socially constructed as it is – to make an effective contribution to pressing real-world environmental issues?

For Haraway, relativism is the mirror twin of totalization. While totalization promises vision from everywhere equally, relativism promises vision from nowhere, or rather, 'relativism is a way of being nowhere while claiming to be everywhere equally' (Haraway, 1991: 191). She argues instead for *situated knowledges* which recognize partiality and positionality and are accountable for their own claims.

> The goal is better accounts of the world, that is, 'science'.
> Above all, rational knowledge does not pretend to disengagement: to be from everywhere and so nowhere, to be free from interpretation, from being represented, to be fully self-contained or fully formalizable. Rational knowledge is a process of ongoing critical interpretation among 'fields' of interpreters and decoders. Rational knowledge is power-sensitive conversation. (Haraway, 1991: 196)

This is not easy. How do we 'have *simultaneously* an account of radical historical contingency for all knowledge claims and knowing subjects, a critical practice for recognizing our own "semiotic technologies" for making meanings, *and* a no-nonsense commitment to faithful accounts of a "real" world' (Haraway, 1991: 187; emphasis in original)? Part of my argument in this book is that scientific and humanities conversations about long-term human interactions with the Earth are now converging in important ways that contribute to 'faithful accounts'.

Scientists trained within an empiricist and positivist tradition often have two further difficulties with much humanities research: they consider that it produces results which are self-evident or a matter of common sense, and they consider it lacking in methodological rigour. As I try to show in this chapter, many of the most important concepts under discussion in this book, including ones central to the environmental issues of Part IV, are understood very differently in different historical and cultural contexts. There is not necessarily any 'common' sense about these issues, and understanding the grounds of commonality and difference is crucial to solving very real environmental problems. Extending this thought, we need also to consider the research process as an arena of cultural practice, and understand how it gives rise to particular conceptualizations of environmental issues as 'self-evident'. This is the theme of Chapter 5.

The methodologies on which researchers referred to here draw are diverse, rigorous and demanding of a diverse range of skills, the latter including both quantitative and qualitative approaches. Anderson and Gale's summary of the requisite qualities gives a sense of both the excitement and the demanding nature of the challenge:

> a sensitivity to people's subjectivity, a keen eye to the topographies that people fashion, a conscientious contextualising of the connections between people and place, a commitment to rigorous documentation and plausible argumentation, and an honest confrontation with our values and the relativity of our own interpretations. (Anderson and Gale, 1992: 11)

Harding argues that it is necessary to live with tensions and ambiguities, but that these in turn have great potential to be creative (1986: 243–4). The concepts of culture, nature and landscape provide excellent examples of such tensions, and they are explored separately below. An important feature of recent debates has been the attempt to reinvigorate old concepts by linking them to contemporary social theory (Box 4.2).

4.1.1 Culture

Recent workers in cultural geography have paid critical attention to the conception of culture inherited from Carl Sauer and the Berkeley school. There have been particular challenges to the so-called 'superorganic' view of culture, in which culture is understood as a total package, with a life of its own operating at a higher level than the individual. If culture can only be explained in its own terms, without reference to specific social forces and contexts, then its explanatory power is severely weakened (Jackson, 1989: 18). It is thus better understood as a process, 'a dynamic mix of

---

Box 4.2 Linking landscape to social theory: common threads in recent debates

- A reformulation of the concept of culture emphasizing human action over passivity. Culture is thought of as a process and expression of negotiated, even contestatory, personal and group interactions, hence it is constantly changing and contingent on context. This contrasts with a traditional superorganic conceptualization;
- an emphasis on the symbolic, as well as on the behavioural interaction or recurrence between humans and their environment that attempts to reconcile the tensions between individual action and cultural structures;
- a problematization of social categories, such as gender, ethnicity, class and race,

and examination of the ways landscape is implicated in the construction and maintenance of these categories;

- the centrality of symbolic expression in the landscape and metaphorical conceptualizations of human–environment interaction as 'text', 'theatre', 'carnival' and 'spectacle' to emphasize the arrangement and manipulation of environments by power structures;
- an awareness of the power of language by subjecting landscape narratives (and authors) to critical reflection and self-conscious interpretation that reveals ideological bias;
- explicit or implicit connections to theoretical frameworks, such as neo-Marxism, post-structuralism and post-modernism.

Source: Rowntree (1996: 140)

symbols, beliefs, languages and practices that people create, not a fixed thing or entity governing humans' (Anderson and Gale, 1992: 3). Within an interdisciplinary conversation, cultural geographers are thus now more likely to recognize a plurality of cultures, and analyse 'the way cultures are produced and reproduced through actual social practices that take place in historically contingent and geographically specific contexts' (Jackson, 1989: 23).

In providing a very brief summary of these trends there is a risk of reifying both contested concepts and complex theoretical positions. There has been considerable debate over whether what is sometimes referred to as the 'new cultural geography' throws the baby out with the bathwater – (*see*, for example, Price and Lewis, 1993; Sluyter, 1997; Willems-Braun, 1997), or paradoxically itself reifies something called 'culture' (D. Mitchell, 1995; Jackson *et al.*, 1996). We should be conscious of both continuities and differences in changing understandings of 'culture'. Such attention to historical process, and the historical contingency of particular situations, is a feature of both Sauerian and new cultural geography, and is particularly relevant to the themes of this book.

To some extent the critique of Sauerian approaches is also an attempt to broaden the subject matter of cultural geography beyond the physical expression of culture in the landscape. Much recent attention has been directed to imaginative and symbolic geographies (Cosgrove and Daniels, 1988; Agnew and Duncan, 1989; Hirsch and O'Hanlon, 1995), often in urban contexts. Further, the idea that culture was the preserve of the exotic – the elite (culture as opera, ballet and art) and the Other (cultures of rural, 'traditional' societies) – was challenged (Anderson and Gale, 1992). For many cultural geographers the everyday geographies of mostly urban, mostly capitalist societies have commanded new attention. Thus the structures of daily life that tend to be taken for granted, or seen as natural, are themselves understood as problematic cultural worlds.

4.1.2 The social construction of nature

Nature is one of the most problematic of these supposedly natural categories. It is also one of the most important, since different conceptualizations of nature tell us much about how human beings understand themselves in relationship to the wider world. The fact that we have the same word in English for 'the essential qualities of a thing' and 'the features and products of the Earth itself, as contrasted with those of human civilization' shows how deeply embedded in Western thinking is the dualism of humans and nature. Yet recent work across several disciplines emphasizes (a) that the dualistic Western conception has been produced and reproduced in quite particular historical circumstances, and (b) that different cultures conceptualize the relationships between humans and their world differently (Merchant, 1980; Evernden, 1992; Simmons, 1993b; Cronon, 1996b). These works have also been influenced by research such as that discussed in Chapters 2 and 3 which shows the pervasiveness of human influences on the Earth. The key tension here is how to frame nature 'as both a real material actor and a socially constructed object' (Demeritt, 1994). In other words, how do we deal simultaneously with nature's autonomous dimensions and human constructions (both mental and material)?

These are not just intensely interesting intellectual questions, but have clear relevance to debates about environmental change, as Proctor and Pincetl's (1996) discussion of spotted owl preservation in the USA demonstrates. They show how conservationist acceptance of a logic of purified nature led to different outcomes for spotted owl protection in the Pacific North-West and in southern California.

The legislative context of these battles is the Endangered Species Act, which itself is argued to exacerbate the distinctions between pure and hybrid organisms and spaces. In the northwest USA the northern spotted owl became a flagship species, using the Endangered Species Act, in campaigns to save old-growth forest. Following this lead, efforts to protect the California spotted owl focused more on the spectacular Sierra Nevada forests than on the drier forests of southern California. 'These efforts have thus neglected owl habitat lying beyond the margins of purified nature, remaining silent about the devastating effects of urbanization on the sustainability of natural systems in the National Forests in southern California' (Proctor and Pincetl, 1996: 702).

In the context of this example, pure nature is also associated with public land, while private land is relegated to the realm of 'culture'. Thus 'the profoundly significant land transformations going on ... on privately owned land ... remain fundamentally unchallenged except on a selective basis' (Proctor and Pincetl, 1996: 701). This type of analysis does not ignore issues of political pragmatism or scientific considerations of habitat quality. Rather, by including a focus on 'the categorical and spatial distinctions between landscapes prioritized for protection and landscapes given less priority or ignored altogether' (p. 686), Proctor and Pincetl situate the case study within a bigger biodiversity picture. They argue that the purificationist approach generates further endangered species by ignoring the fundamental driving force that leads to species extinction, namely development on private land.

The agency of nature has been an important theme within the relatively new subdiscipline of environmental history. While the agency of nature would not surprise anyone working on long-term environmental change, it was relatively neglected within social theory until the past decade (FitzSimmons and Goodman, 1998: 196). This work is most developed in the USA, through the writings of Merchant (1989), Cronon (1983, 1991) and Worster (1985). Two important features of this discourse are its recognition of nature as being present and active in human affairs, and of reciprocity between nature and society (FitzSimmons and Goodman, 1998). This is not to suggest a uniformity of approaches in environmental history; indeed, FitzSimmons and Goodman outline several disputes among practitioners, particularly with regard to how different categorizations of nature are reified.

We can only wonder why the same terrain has not been engaged with so visibly in recent historical geography (although note, for example, the works of Powell and Meinig (Powell, 1977, 1988a, b; Meinig, 1962). It is

possible that geographers have been battling the ghosts of environmental determinism for so long that they lack a conceptual vocabulary for an active but non-determining nature. Within human geography, discussions of nature and landscape have cohered around the social construction of meaning, rather than the biophysical environment itself.

4.1.3 Landscape

Like culture, landscape is no longer (if it ever was) something that can be simply defined. Rather, it is a concept whose problematic status makes it interesting.

Just as in the analysis of cultures, there are a number of threads in current conceptualizations of landscape. From something originally understood mostly in physical terms, there has been a broadening of interest into the social dimensions of landscape. Extensive literatures are thus found not only in geography, but increasingly in anthropology and archaeology (Bender, 1993a; Head *et al.*, 1994; Hirsch and O'Hanlon, 1995; Ucko and Layton, 1999).

Of particular relevance to the concerns of this book is interest in the interaction between different dimensions of landscape: physical, social and symbolic. In identifying those different dimensions, we need to be wary of reifying them as economic base, social structure and symbolic or sacred superstructure. Instead I try to draw on studies where the interplay between and interpenetration of the dimensions are explored.

The multiplicity of meanings of landscape could be usefully ambiguous, or it could be just plain frustrating (Gosden and Head, 1994; Rowntree, 1996). There is a sense from contemporary everyday usage in English that people have little difficulty with multiple, including metaphorical, meanings of landscape. We speak quite comfortably, for example, of 'political landscapes', of 'moral terrains', of 'mapping out strategies'. In the same spirit we can summarize the main uses of landscape, as identified by Rowntree (1996), for human geography (Box 4.3), without attempting to focus on one or other (Illustration 4.1).

Box 4.3   Landscape themes in environmental change

Landscape as ecological artefact

- This is the theme of the Earth as transformed by human action, discussed in terms of the long-term record in Chapter 2. It is also seen in recent environmental history.

Landscape as evidence for origins and diffusion

- Landscape primarily as a backdrop for developments inferred or deduced from other sources, e.g. Sauer's classic study on agricultural origins.

Landscape as material culture

- Fundamental emphasis on the 'look of the land', e.g. barn types, fence architecture, field patterns.

Urban landscapes

- A range of questions now being asked about cityscapes, e.g. ethnic constructions of space and place, shopping malls as landscapes of myth

Art, literature and landscape meaning

- Long-standing interest in the ways people depict the landscape in various media. These feed back into society by privileging certain scenes or ways of seeing.

Landscape as visual resource

- Visual quality as part of a larger environmental quality, e.g. visual blight as equivalent to pollution.

Landscape as ideology

- Landscape studied for ideas and objectives that act as political social guidelines for a national culture

Landscape's role in the production and maintenance of social categories

- How individuals and institutions shape and control the environment to produce and reinforce power relations, especially of race, class and status

Landscape as text, symbols and signs

- Landscapes treated metaphorically as texts that were authored and, hence, could be read by insightful observers. Interaction between reader and text becomes as important as the material objects themselves.

Source: Based on Rowntree (1996)

Cultural landscape will remain a fundamental concept in human geography more because of its momentum than because of its conceptual clarity. *Landscape's* definitional or methodological shortcomings notwithstanding, the term has been used too long and in too many ways to be radically revised. Furthermore, a receptive lay audience awaits our writings about *landscape* because it strikes intuitive chords about the visual and material environment. (Rowntree, 1996: 147; emphasis in original)

The multiplicity of landscape also has political implications. In Australia the incorporation of notions of landscape into contemporary discourse has occurred partly in the context of Aboriginal claims to land. Settler Australians have been confronted with the very different conceptualization of land, and of human relations to land, expressed by Aboriginal people. They illustrate well how the same physical landscape can be conceptualized quite differently by different cultural groups. For example, pastoralists would think in terms of its suitability or otherwise for cattle grazing; a tourist might focus on its frontier or wilderness dimensions; a palaeoecologist might read it as an archive of long-term environmental change. It would be inaccurate, however, to suggest that so-called Dreaming stories are purely representational or symbolic devices in Aboriginal life. Rather, having become aware of different readings of landscape, we should extend our thinking to consider the ways in which the symbolic and the material dimensions penetrate one another, for example when Aboriginal people refer to their grandfather as a 'bush plum' (Young, 1992), or argue that uranium mining should not occur on particular land because it is 'sickness country'.

ILLUSTRATION 4.1 Pagoda, Namsan Mountains park, near Kyongju, South Korea. This landscape is one that can be understood as simultaneously transformed by human activity and symbolic of social meanings (by permission of Gordon Waitt)

Although the particulars vary, the notion of contested landscapes has resonance in many parts of the world. Contexts include the contest between settler and indigenous, in for example Canada and New Zealand (Pawson, 1992;

Young, 1992), and landscapes of exclusion, in which certain categories of people are constructed as being out of place (Sibley, 1992; Bender, 1993b). The process of naming places is an important part of claiming (and reclaiming) space in such contests (Berg and Kearns, 1996). As we will see in the parts of this book that deal with management of cultural landscapes, one way of gaining power over land is to have one's own definition – of culture, nature, landscape, or all three – appear as the commonsense one, the natural one.

4.2 Wild and tame

Nowhere has the rethinking of culture, nature and landscape converged as strongly as in the recent reassessment of the wilderness ideal. Definitions of wilderness have a long history of change (Nash, 1967; Tuan, 1971; Powell, 1977), but I refer here particularly to the romantic wilderness ideal that has underpinned conservation and national parks policy in frontier societies such as the USA and Australia over the past century. The challenge to wilderness has come from diverse lines of evidence, including palaeoecological and archaeological demonstrations of long histories of human occupation (Head, 1990; Denevan, 1992; McNiven and Russell, 1995; Cronon, 1996a) and indigenous voices for whom wilderness areas are home (Sultan, 1991; Langton, 1995–6, 1996), as well as the general critique of nature discussed above (Cronon, 1996b). The discussion is cross-cut by an increasing awareness of the colonial heritage in the supposedly post-colonial environmental thinking and policies of New World societies (Head, 2000). As we discuss in later chapters, wilderness is less of an issue in areas where the long history of human occupation is more widely recognized.

Wilderness is only the most visible of a series of issues swept into this critique. Unsettling the boundaries between human and nature, tame and wild (Anderson, 1997), civilized and savage, native and introduced, challenges many of the conceptual and physical boundaries previously taken for granted in environmental management. What is the place of people in national parks? What is the status of feral animals or endangered species (Whatmore and Thorne, 1998)? Fundamentally, what is a natural environment?

There are those, particularly conservationists working within a natural science framework, who see the critique as fiddling while Rome burns, as being 'just as destructive to nature as bulldozers and chainsaws' (Soulé and Lease, 1995). The problem with this is the refusal to see science itself as a social process. For Soulé (1995) there are at least nine distinct cognitive formations of nature in Western thought (Magna Mater, Unpredictable and Evil Bully, Ageing and Reluctant Provider, Wild Kingdom, Open-Air Gymnasium, New Age Temple, Wild Other or Divine Chaos, Gaia, Biodiversity). Biodiversity is 'the living nature of the contemporary Western biologist' (p. 140); science is seen as either equivalent to, or at the very least a transparent window on, living nature. The irony of the rejection of the critique

> by some natural scientists and others, is that it is predicated on a particular social construction of nature – one which is purified of its embed-dedness in cultural schemes of knowledge and transformative practices, and hence stakes out this pure nature as worthy of protection from adverse human influence. (Proctor and Pincetl, 1996: 685)

Good science pays attention to the social context of its own production. We shall discuss this in detail in Chapter 5.

4.3 Landscapes of the prehistoric past

In this section we look at different ways in which the notion of the social landscape is being used in archaeology. Archaeology provides a particularly useful disciplinary example because it illustrates a number of the general themes already raised:

- There are multiple meanings of the notions of social or cultural landscape.
- A dynamic is created between humanities questions and the use of natural science techniques to find answers.

- Further, since archaeology by definition deals with the material record of human behaviour, it is well placed to analyse the relationships between materiality and the less visible elements of social and symbolic activity.
- The long time perspective of archaeology matches that of this book.
- In relation to cultural heritage issues discussed in Part IV, archaeology reminds us that valuing and remembering the past is not a recent process, but can be shown to have existed at various times in prehistory. Landscapes of memory are not neat and tidy landscapes, but are inscribed and rein-scribed with meaning in a patchwork of ways.
- Constructing meaning about the categories of nature and culture, and the boundaries between them, is part of long-term human history.

The temporal perspective brings with it dilemmas of appropriate scales of analysis. One of the problems in archaeology has been the tendency for environmental variables to be viewed as long-term processes and constraints, and social variables as short-term ones – for the human story to be in effect a story of the gradual enculturation of nature. I disagree with that view, and want to put forward analyses in which not only are the categories culture and nature problematized, but also both social and environmental processes, however we constitute them, interact over a range of timescales. However far back in time we go, people were acting according to their social construction of their world. Projecting into the future, we have not overcome our embed-dedness in the physical world. If we had, there would be no 'environmental problems' and no reason for this book. The exciting opportunity here is to examine interactions between humans and their landscapes over very long timescales. Those changes were not, however, predetermined or inevitable.

The tradition of landscape archaeology that I focus on here is most strongly developed in north-west Europe. It is important to distin-guish it from the North American tradition whose themes, to overgeneralize, have more in common with the idea of 'landscape as artefact', such as we saw in Chapter 2. Both traditions utilize a range of scientific evidence within archaeology, such as palaeoecological indicators, radiometric dating and analysis of use wear and residue analysis on stone tools. However, it can be argued that North American landscape archaeology is characterized by a strongly scientific methodology (e.g. Rossignol and Wandsnider, 1992), whereas there is more interest in Britain and north-west Europe in embodied and phenomenological approaches to landscape (Bender, 1999). In the former, workers are more likely to align their approaches with those of Carl Sauer, for example with respect to the need for multidisciplinary field studies (Dunning *et al.*, 1999), or in relation to the changing nature of cultural landscapes (Gartner, 1999). They are also more likely to still feel a need, in working with palaeoecological evidence, to differentiate themselves from environmentally deterministic thinking (Erickson, 1999). In Britain and north-west Europe there is a trend towards thinking about how different people live in and move around sites and landscapes, creating, resisting and reworking meanings. As Cooney emphasizes,

> we cannot hope to think like a prehistoric person did about their landscape but we can reconstruct an overview of what the elements of that landscape may have been and then try to understand what they meant for the people who were carrying this landscape round in their heads. (1999: 47)

The subject matter at issue in this literature is the Mesolithic–Neolithic transition in northwest Europe, particularly Britain, and the nature of the Neolithic itself (for overviews and bibliographies, *see*, for example, Thomas, 1991, 1996; Barrett, 1994; Whittle, 1996; Van Gijn and Zvelebil, 1997; Edmonds and Richards, 1998). Three themes are highlighted here: the multiple notions of landscape; the notion of domestication as a state of mind rather than as a genetic transformation of plants and animals; and the idea of remembering in the past.

There is perhaps a risk here in choosing the most Eurocentric of possible examples. Have not Western understandings of landscape and the past been most fundamentally challenged by non-Western and indigenous voices (Bond and Gilliam, 1994; Schmidt and Patterson, 1995)? Those voices have certainly been important in contributing to the climate of challenge in which social and cultural processes have come to the fore, and they are explored in more detail elsewhere in this book. However, for the purpose of illustrating the multiple ways in which social landscape can be theorized within archaeology, I have found this particular European debate to be the most useful.

Analysis of the research process itself is critical to these debates; for example, we will see how the boundary between the Mesolithic and Neolithic was reified by different approaches and traditions of scholarship. While the new approaches focus on social rather than ecological processes, this is not to suggest that palaeoecological evidence is irrelevant in these debates. Indeed, as we saw in Chapter 2, some of it has been used as part of the critique of traditional models (Brown, 1997a). This section should thus be read against the different perspectives presented in section 2.3 of Chapter 2.

4.3.1 Multiple landscapes

Cooney's (1999) overview of social landscapes in Irish prehistory provides a useful example of multiple landscapes and their attendant methodological difficulties. He begins by using the complexity of contemporary perceptions of the Irish landscape as a reminder of the difficulties in interpreting the prehistoric past. The tourist image of Ireland is 'an uncrowded, green and pleasant land, a timeless land where visitors from more frenetic lifestyles and landscapes can come for rest and relaxation' (p. 46). This is argued to have links to both colonial (empty and awaiting development) and romantic (wild, uncultured and allied to the eighteenth-century invention of 'the Celt') views of the Irish landscape. A combination of such images is commonly utilized in the development of national identity. These themes will be explored further in Part IV when we consider the complexity of landscape perceptions in tourism and cultural heritage management. Difficult though such complexity may make the interpretation of the prehistoric record, it also emphasizes that all landscapes are essentially social, in that 'we perceive, understand and create the landscape around us through the filter of our social and cultural background and milieu' (p. 46).

Cooney's study (1999) uses five dimensions of prehistoric landscape: 'landscapes from the outside', 'people, pathways and places', 'sense of place', 'landscape as context of activity', and 'transforming the landscape' (cf. Box 4.3). I summarize here the main features of each one in order to illustrate how a multi-layered interpretation can be built up. His central argument about the Irish Neolithic is that

> a sense of place can be seen as a major theme running throughout the period, not only in the sacred sphere but also in the secular realm. The definition of space, or the fixture of elements within it, seems primarily to have been intended to create a sense of permanency in the relationship between the community and the landscape. A concern with inserting and asserting a long-term human presence in the landscape as part of the natural order was an aspect of both secular and sacred life (Cooney and Grogan, 1998).

Landscapes from the outside refers to the extent of prehistoric settlement as discerned from the spatial distribution of archaeological evidence such as monuments and artefacts. It also draws on palynological evidence of human impacts on vegetation, as is discussed in different contexts in Chapter 2, although in the Irish context Cooney argues that circular arguments have been used to reinforce interpretations of changing settlement intensity.

The dots-on-a-map approach to settlement patterns can lead to static depictions of complex systems in which there is little consideration of *people, pathways and places*. As well as construction and use of megalithic tombs, Neolithic Ireland (since the fourth millennium BC) shows evidence of field boundaries, long-term forest clearance for agriculture, and substantial houses. Cooney interprets this as indicating that 'the house was at the core of settlement and life; the areas used for farming, gathering and hunting on a daily basis would

have been the most familiar parts of the landscape' (1999: 50). Sites visited less often, such as stone quarry sites or cemeteries, would also have been seen as important, but perceived in quite different ways. Cooney contrasts the Irish evidence for early sedentism with the later retention of mobility in Britain. Archaeologists in Britain tend now to view the Mesolithic and early Neolithic landscape 'as one in which mobility was the binding thread, linking places of social and religious significance' (Cooney 1999: 49), with a place-bound landscape perspective centred on the household not developing until the second millennium BC.

A sense of place is developed through long-term interactions with permanent structures such as megalithic tombs. For example, at Loughcrew an elongated ridge divides two river systems. Clusters of passage tombs on the ridge crests surround large, focal hilltop cairns, which oral tradition relates were dropped by a witch (Cooney, 1999: 52–4) (Figure 4.1). Similarly, Silbury Hill in southern England is visible from many places in the surrounding countryside (Figure 4.2). The focal point this affords is interpreted by Barrett (1994: 31) as making possible the 'ceremonial presencing of an elite' who were both visible and unreachable.

As Barrett argues, most archaeological approaches 'privilege the date of creation over the chronology of appreciation' (1999: 22). In fact, later people can interact with long-lived landscape features in ways quite different from the intent of their creators. The creators may also have never completely realized their own intent. Barrett (1994: 13) argues, for example, that the enclosure and stone circles at Avebury, linked visually to Silbury Hill, are 'the physical remnant of a number of abandoned projects and not the culmination of a series of planned phases' (Illustration 4.2). Nor is this point only or particularly relevant to structures created by people. Hunter-gatherers become intimate with landscapes through somewhat different patterns of mobility; investment of landscape features with meaning is characteristic of such societies. For example, the most permanent and prominent parts of 'unmodified' landscapes, such as rock outcrops, are often particularly

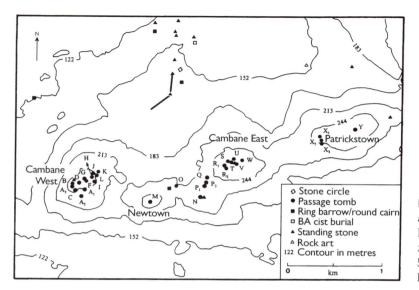
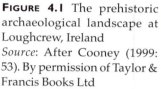

◇ Stone circle
● Passage tomb
■ Ring barrow/round cairn
▫ BA cist burial
▲ Standing stone
△ Rock art
122 Contour in metres

0   km   1

FIGURE 4.1 The prehistoric archaeological landscape at Loughcrew, Ireland
Source: After Cooney (1999: 53). By permission of Taylor & Francis Books Ltd

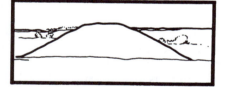

FIGURE 4.2 Four views of Silbury Hill showing how the platform projects above the skyline
Source: After Barrett (1994: 31). By permission of John Barrett

ILLUSTRATION 4.2 Enclosure, ditch and stone circles at Avebury, England, with interpretive signage

significant places in Aboriginal Australia (Taçon, 1991; Fullagar and Head, 1999).

The discussion above focuses on the attributed meanings of relatively spectacular features, but of course landscape also provides the *context of activity*. In Ireland this raises questions about the relationships and archaeological visibility of day-to-day and ceremonial or ritual activities (the profane and the sacred). This is tied into the patterns of use of wetland and dryland areas, for example the discussion about whether enclosure sites with materials deposited in nearby water bodies are best interpreted as frontier fortifications or as sacred sites with wetland elements (Cooney, 1999: 58–9).

Transforming the landscape is, then, understood not just as large-scale physical transformations and emplacement of structures, although these are important (e.g. Illustration 4.3). For Cooney, this theme represents 'continuity within change' (1999: 60) – continuity being the place, which is experienced and remembered in multivalent and ambiguous ways.

4.3.2 Domestication as a state of mind

The transition from hunter-gatherer to agricultural society is usually related to the domestication of plants and/or animals, the beginnings of cultivation and the onset of sedentism, not necessarily in that order. Within a natural science tradition, domestication is defined by the morphological changes that plants, particularly, undergo (e.g. Roberts, 1989: 96). There is increasing attention in the humanities to the notions of domestication as a state of mind (Bradley, 1997), and the domestication of the environment (Yen, 1989). The significance of the transition is exacerbated by the different traditions of scholarship on either side of the threshold. 'Mesolithic specialists emphasize adaptation to the natural environment, whilst those who study the Neolithic are more often concerned with ideology and social relations' (Bradley, 1997: 13). Thus 'in the literature as a whole, successful farmers have social relations with one another, while hunter-gatherers have ecological relations with hazelnuts' (Bradley, 1984: 11).

However, the transition is being rethought on the basis of both empirical and theoretical evidence. While there is still heated debate about many of these issues, three trends are highlighted here. One is the regional variability which emerges when smaller scales of analysis are used. Second, empirical evidence also breaks down the idea of a clear temporal boundary – showing, for example, intensive use of plants in the Mesolithic (Zvelebil, 1994) and late survival of mobility in Britain. These contribute to a narrative that is more fragmentary in space and time, and show that settlement, subsistence and ceremonial changes need not occur together, nor in predetermined directions. Thus the empirical

Illustration 4.3 V-shaped fish weir built from basalt boulders, Lake Condah, south-eastern Australia. During flooding of the lake, weirs and channels were used by Aboriginal people to direct fish and eels into holding ponds and traps

evidence increasingly illustrates the limitations of deterministic evolutionary overviews; these are histories 'framed by contingency rather than by inevitable process' (Whittle, 1996: 369).

Third, the social emphasis in interpretation – itself not a monolithic entity – occurs in such a context. Some social explanations see the Mesolithic–Neolithic transition as significant, but consider it first a matter of changing world-views rather than simply a subsistence change. Others emphasize continuities between hunter-gatherer and agricultural ways of seeing. The authors cited in this section would disagree about many aspects of the transition. It is not my intention to review in detail, much less attempt to resolve, these complex debates. Rather, in looking at examples – albeit perhaps contradictory ones – of how domestication is being reconceptualized in terms of its social dimensions, I hope to show how this extends to cultural landscape studies generally.

In Chapter 2 we saw considerable variability in the palaeoecological evidence for associated environmental transformations in different parts of the world. Archaeological evidence has shown what Gosden refers to as three Neolithics even within the Near East / European sphere, let alone the Far East, New Guinea or Mesoamerican variations.

> The Near East and south-east Europe built villages, often of mud brick, and these relatively dense communities struggled with tensions between the genders and the fear of death. Central Europeans built large longhouses in clearings in the forests and emphasized transitions between the internal domesticity of the house and external natural forces contained within the forest. Western and northern European communities adopted the novel resources of farming late and incorporated these gradually within traditional Mesolithic practices. The major effect of farming was to extend the range of communities, from a concentration on coastal marine resources to culti-vation of plots inland. In these areas there was no sudden adoption of sedentism, but a gradual reduction of mobility over a millennium or more as the garden plots became increasingly important to life. (Gosden, 1994: 159)

For Hodder (1990) this variability is under-pinned by a shared symbolic and social process in which the culture / nature duality is a central opposition. Separation and control of the wild,

agrios, is the means of accruing social and cultural prestige. This process begins in the Palaeolithic with the bringing of aspects of the wild, such as fire, into the cultural domain (pp. 287–9). Categories such as 'death' or 'fire', Hodder emphasizes, are not inherently wild, but are categorized and defined as such by social processes, the particular expression of which varies in space and time. By the Neolithic the *domus*, the cultural, becomes asso-ciated with the house and the hearth as both symbolic focus and focus of domestic production. The development of agriculture is thus understood as a conceptual 'culturing of the wild' (Hodder, 1990: 86).

In contrast, Bradley reads the Mesolithic archaeological record as showing the natural world 'as a creative principle rather than a source of danger' (1997: 16). However, social and conceptual factors are also to the fore in this framework. In his elaboration of domesti-cation as a state of mind, Bradley shows that Mesolithic ritual in Europe emphasized the natural world through features such as use of red ochre, deposition of antlers with the dead, and food remains in funerary assemblages. The gradual acceptance of ownership of resources and the building of monuments is related, then, to a breakdown in such thinking and the adoption of new beliefs. This also may help to explain why, far from the adoption of farming being an automatic process, some hunter-gath-erers resisted it:

> In a world in which human identity was not felt to lie outside nature – a world in which natural places could take on a special significance – monuments would have little part to play ... Until that belief system lost its force, domesti-cation may have been literally unthinkable. (Bradley, 1997: 16)

In a related but again quite different argument, Sherratt (1995: 248) argues that megaliths were 'a set of monumentalised messages', part of the 'cultural rhetoric' by which Neolithic groups sought to convert indigenous foragers, concen-trated in coastal areas of north-west Europe, to a Neolithic way of life. What these workers share, then, is a thinking about farming as 'a total package of economic, social and ideo-logical practices ... The phenomenon of megalithism is associated not with a particular kind of economy, but rather with a social process in space' (Sherratt, 1995: 257, 258).

In using the megalithic example it is not my intention to imply that the most salient or visible social landscapes are sacred ones. Nor do I want to perpetuate the notion of an ideological superstructure on an economic base. Indeed, I have tried to refer to work which challenges the basis and separation of those very categories. Nevertheless, it is a fact of archaeological life that the debates of interest here are most developed in the context of a landscape dominated by long-lived symbolic monuments. As Barrett reminds us, the considerable time and labour expended in ritual and ceremonial structures

> was invested in long-term projects which extended over some two to three millennia and the investment must have been only a fraction of the total energy reserves demanded by more mundane, everyday activities ... When they are expressed as a fraction of the lived allocation of time and space, and when they are set against the more densely occupied background of cultivation plots, pasturage and woodland, then ritual and ceremonial activities begin to appear almost marginal. (Barrett, 1994: 132)

It is important to stress that these everyday landscapes are just as social, even if the outcome of habitual activity or 'unthought practice'. 'Landscapes are social products, but are not first and foremost symbolic constructs or landscapes of the mind' (Gosden, 1994: 81). A number of different examples could be included here. Gosden (1989) characterizes Melanesia as a 'landscape of debt', in which the demands of reciprocal gift exchange have resulted in a distinctive settlement pattern and sites for specialized production. Habitual patterns of inter-island trade in the Arawe islands, Papua New Guinea, stimulated Gosden and Pavlides to put forward the notion of seascape in interaction with landscape (Gosden and Pavlides, 1994). Landscapes of many Polynesian islands can be understood as landscapes of 'chiefly privilege', with investments in irrigation and places of elite residence (Bayliss-Smith and Golson, 1999).

The differential preservation of archaeological remains is influential here. The persistence and durability of monuments in the landscape leads to several biases, including foci on landscapes of prestige rather than everyday landscapes, and on male wealth and power (Mulk and Bayliss-Smith, 1999) (Figure 4.3a, b).

There is an important counterpoint to this discussion in studies of hunter-gatherer worldviews and their relationship to patterns of land use. Discussion of the relevance or otherwise of the Mesolithic–Neolithic boundary in Europe is paralleled by the spatial and conceptual boundary between Australian hunter-gatherers and New Guinea agriculturalists (Gosden and Head, 1999). Rather than a clear dichotomy and parallel patterns of subsistence, attachment to land and ceremony, considerable variability is indicated. Aboriginal groups along the River Murray, in south-eastern Australia, used cemeteries 'as one of the symbols validating corporate ownership of that territory' (Pardoe, 1988: 14). Fire was used by hunter-gatherers not just for economic purposes but to bring landscapes into the human realm (Head, 1994b). New Guinea highlanders express attachment to land in terms of lines of power (Ballard, 1994) which have similarities to Aboriginal Dreaming tracks.

The idea that a particularly integrated worldview may have been the reason for the non-adoption of agriculture in certain areas has also been explored by Chase (1989). In an important contribution to a broader conceptualization of domestication, Chase distinguished between *domus* (a hearth-based area of exploitation), *domiculture* (localized packages of interaction between people and resources) and *domain* (following anthropologist Stanner's work, the geographical area of legitimized use in Aboriginal society). This built on a study of Aboriginal–plant interactions in Cape York Peninsula (Hynes and Chase, 1982), in which people's conceptual stance towards plants is intimately related to their impacts on the vegetative landscape.

There are many problems and ongoing questions raised by these studies. What are the methodological issues involved in reading people's conceptualization of the world from the biased and fragmentary evidence of the archaeological record? In rendering domesti-

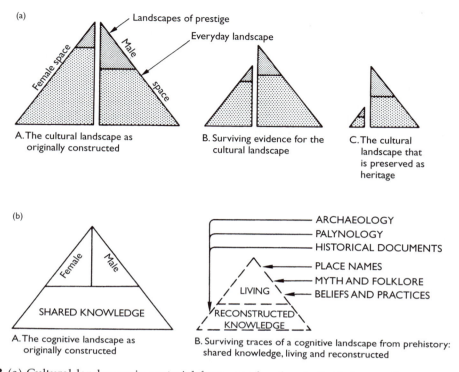

FIGURE 4.3 (a) Cultural landscape in material form, gendered and divided according to its prestige or everyday purpose. (b) The cognitive landscape, as originally constructed and as surviving or reconstructed *Source*: Mulk and Bayliss-Smith (1999: figures 24.2, 24.3). By permission of Taylor & Francis Books Ltd

cation as a social as well as an ecological process, are researchers not simply reinforcing dualisms of culture/nature and tame/wild rather than considering that prehistoric people may have conceptualized things in non-dualistic ways? To what extent does domestication as socially conceived simply reinforce its determining role in the trajectory of human history? We cannot answer those questions here but it is important to acknowledge them.

Much of this chapter, indeed much of this book, is about the past. The long-term perspective has provided important insights. Cultural landscapes are not the preserve of sedentary or urban societies. People have attached meaning to their landscapes, and shaped them both consciously and unconsciously, throughout human history. Those meanings have been contested, and have included values placed on remembering the

past. Gosden and Lock comment on the reluctance of archaeologists to ascribe historical consciousness to the peoples being studied, and illustrate that consciousness for the prehistoric people living around the Ridgeway in south Oxfordshire, southern England (Gosden and Lock, 1998). The White Horse at Uffington, one of the chalk-cut figures of southern England, is possibly as old as 3000 years (Illustration 4.4). Chalk figures disappear if they are not cleaned every 5–10 years; the White Horse is argued to have had some historical power in the landscape throughout a very long period, although the social context of that power may have changed considerably (Gosden and Lock, 1998). The evidence we now have of those pasts is fragmentary and our interpretation of it problematic; close attention to the production of knowledge about the past is required. In Part IV we will consider the

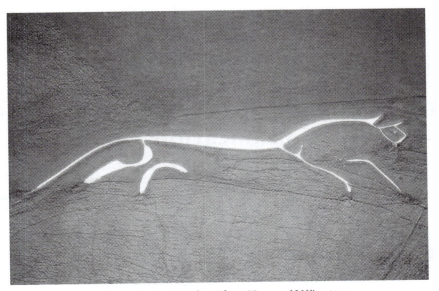

ILLUSTRATION 4.4 The White Horse of Uffington
(by permission of Atmosphere Postcards)

implications of the transformation of these patchwork pasts into heritage to be valued in the present and protected in the future.

However, none of that should be read as implying that cultural landscapes are only about the past. Rather, it gives us some tools for critical self-consciousness about the landscapes being shaped today. For both prehistoric people and ourselves, 'in daily life there is always a tension between the inheritance from the past, the intentions of the present and the possibilities held by the future' (Gosden and Lock, 1998: 4). Some examples of these 'new' landscapes, their biophysical creation and the attribution of meanings are discussed in the second half of the book. The question of knowledge production about the future is a topic of the next chapter.

Production of knowledge and its policy implications

5.0 Chapter summary

This chapter examines the production of knowledge about long-term environmental change and human interactions, and considers the policy implications of particular understandings. It starts with the conceptualization of culture in global change research, focusing on two elements of Proctor's (1998) critique; first, the myth that culture is *separable* from other human dimensions and can be analysed independently; second, the myth of *externality*, that research is not itself a cultural process. The example of climate change illustrates several issues in the cultural production of knowledge: internationalization of research and the use of analogy in applying the long-term record. Two different critiques of the narratives of deforestation from Africa and India are used to illustrate both the production of knowledge and the policy implications. A final example takes the discourse of cultural landscape itself, and examines its application in Norwegian landscape administration.

5.1 Researching culture and the culture of research

How do we produce knowledge about long-term environmental change and human interactions? Since the periods of greatest interest are the deep past, thence projecting towards the future, there is no ultimate test for many competing hypotheses. There will always be heavy reliance on analogical reasoning. This is not to say there is no place for deductive approaches; indeed, in Chapter 2 we saw a number of examples of them in use. I also argued that the trend within much palaeoecological research towards use of multiple scales of analysis, in both space and time, is increasing the application of deductive approaches.

Whether through induction or deduction, and usually through a combination, we construct stories about the past, about the present state of the world environment when seen within a long-term perspective, and about possible futures. In this chapter we shall focus on the process of that story-telling, although not to argue that it should not be happening or that there is a single correct story. Rather, we want to understand something about science as a social and cultural process, and the implications that has for our understanding of environmental change. The examples used here are mostly from the 'global environmental change' literature, which looks most at the present and projected futures. We will also look for points of connection with the global environmental past, although there is not a comparable body of critique in that direction.

Two examples are used to analyse this production of knowledge: climate change and forests.

Research into environmental change and human dimensions is having a variety of inputs into policy. It is one thing for academic researchers to recognize cultural landscapes as products of a complex interplay between social process and materiality, as having shifting and multiple meanings, and as being reworked over time. To what extent can government bodies charged with the management of such landscapes deal with this dynamism and complexity? The first requirement is a recognition that any cultural landscape is a discourse materialized (Schein, 1997). We can then go on to explore 'how discourse matters in the reconstitution of material ecologies' (Robbins, 1998: 69).

5.2 Theorizing cultural processes in global change research

It is hard to imagine a project in which it is more important to have the best possible understandings – both theoretical and empirical – of both 'culture' and the 'environment' than global environmental change. Yet Proctor (1998) argues that culture as conceptualized in most research into the human dimensions of global environmental change is a vague concept offering no analytical clarity. He critiques three mistaken assumptions about culture found in this research (p. 228):

- *separability*: that culture can be disentangled from and analysed independently of other human dimensions;
- *methodological individualism*: that culture is best understood at the level of the individual person;
- *externality*: that research is not itself a cultural process.

In Chapter 4 we discussed the recent retheorization of cultural processes in terms of meanings shared between various, often overlapping, human groups. While not ignoring the individual person, this work shows many scales of agency, up to global ones. For Proctor,

the operative question should not be 'how important culture is relative to other dimensions, but rather: what kinds of shared meanings are connected with the full range of human practices associated with global environmental change?' (Proctor, 1998: 239). In this chapter we shall revisit briefly the notion of separability, and then consider the research process itself in some detail.

Proctor reviewed three major research agenda statements on the human dimensions of global environmental change to examine their engagement with culture. The US National Research Council study illustrates, for example, the notion of separability. It presents culture, in terms primarily of attitudes and beliefs, as one of five human driving forces that can be analysed separately from other human dimensions (Figure 5.1a). Drawing on the critical rethinking of culture that we discussed in Chapter 4, Proctor argues instead that we should understand 'that cultural processes of meaning are implicated in all relevant human practices' (1998: 239), as illustrated in Figure 5.1b. (Proctor does not necessarily give primacy to cultural process alone, noting that other overarching contexts, such as political factors, could also apply to all 'human driving forces' (p. 232). Rather, it is the analytical separation of culture that he is challenging.) There are important parallels between the issues of separating culture from other human driving forces, and separating humans from other parts of ecosystems, as discussed in Chapter 3.

An important dimension of culture is the research process itself. Indeed, in the environmental change arena, science has been crucial to how we conceptualize change and construct problems. We could not even imagine, for example, an ozone hole over Antarctica without drawing on an array of scientific research dependent on high technology. General circulation models (GCMs), the basis of future climate and impacts scenarios, are complex representations of global climate systems possible because of the ongoing revolution in information processing, integrating empirical measurements and remotely sensed information. In projections both forward and backward in time, we are becoming used to

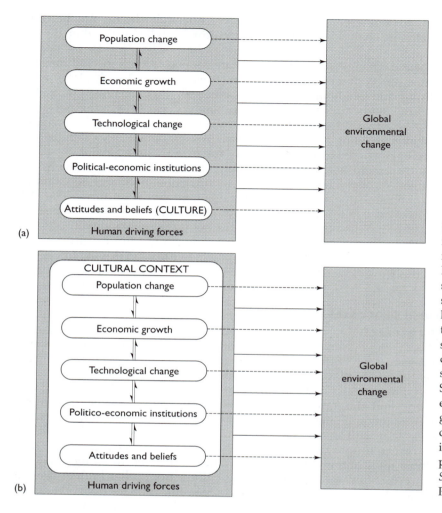

FIGURE 5.1 Human driving forces of global environmental change noted in US National Research Council study, with (a) culture separated analytically, and linked primarily to attitudes and beliefs. Proctor suggests instead (b), all five driving forces involve a significant cultural context. Solid arrows denote net effects of driving forces on global environmental change; dashed arrows indicate possible independent effects

Source: Adapted from Proctor (1998: figure 1)

more pixellated views of the Earth (e.g. Figure 5.2). There is much argument about the accuracy and validity of various reconstructions, models and predictions. There is less consideration of the ways in which the models are themselves cultural constructs, and thus implicated in the transfer of meanings about the environment.

> It is not that the scientific models and ensuing knowledge are empty of culture and politics, but that they are impregnated with them without even recognizing it, let alone the implications. Existing cultural and institutional structures are by default taken as immanent and natural. (Shackley and Wynne, 1995: 124)

Not only the scientific research into environmental change, but even the human dimensions research sees itself as external to the objects of inquiry (Proctor, 1998) (Figure 5.3a).

> Thus, when culturally based attitudes, beliefs, and so forth are mentioned as important filters in the ways people perceive and respond to global environmental change, there is no concession that the cultural filters of the analyst or the analyst's scientific tradition may play any role. (Proctor, 1998: 234)

An important part of good research, then, is to consider its positionality, to consider it as part of the object of inquiry (Figure 5.3b).

How can science best understand its own cultural baggage, and the influence of that

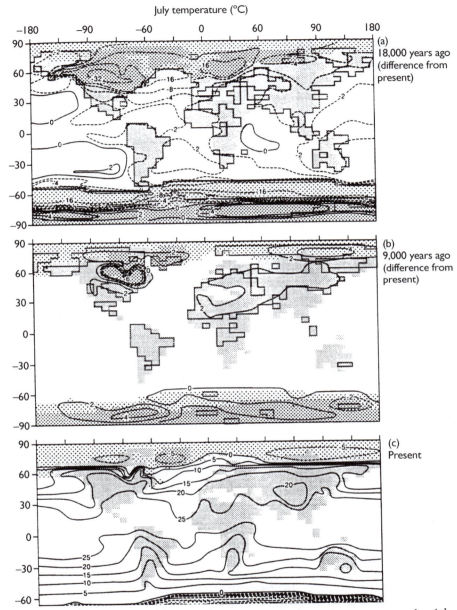

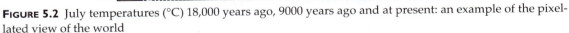

FIGURE 5.2 July temperatures (°C) 18,000 years ago, 9000 years ago and at present: an example of the pixel-lated view of the world
Source: Huggett (1997: figure 4.4, p. 105). By permission of Taylor & Francis Books Ltd

baggage on the articulation and solving of problems? Further to the point made in Chapter 4 about the importance of rigorous attention to the conditions of knowledge production, the next two sections consider the issues of climate change and forest policy within the frameworks of the social construction of scientific knowledge (Demeritt, 1996, 1998; Wynne, 1994).

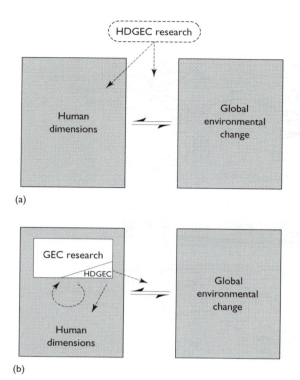

(a)

(b)

FIGURE 5.3 Human dimensions of global environmental change (HDGEC) research as (a) external to the objects of inquiry (noted by dotted arrows), and (b) included as an object of inquiry in the context of global environmental change research
Source: Adapted from Proctor (1998: figure 3, p. 235)

5.3 Climate change

The aspect of global environmental change most firmly lodged in the imaginations of many in the West, including policy-makers, is the issue of climate change, in particular whether industrial activity is exacerbating a trend towards global warming. Modelling and prediction are centred on present and future timescales. However, the deep past section of the long-term record has been mined to provide indications of what the 'normal' rate and directions of change might be, and to provide analogies for a variety of future scenarios. Within this cultural and intellectual climate, apparently arcane debates such as

whether the Antarctic ice sheet melted 3 million years ago take on a heightened degree of relevance (Sugden, 1996). Conscious of enhanced funding opportunities, as well as genuinely believing in its importance, many palaeoecologists have positioned their recent research in terms of providing a long-term and/or analogical perspective on contemporary and future problems (e.g. Bishop, 1988), but rarely in a way that is explicitly articulated. A collection of overlapping discourses can be identified in expert and public debates over climate change (Thompson and Rayner, 1998). In this example we look at two aspects, the internationalization of science and the use of analogues. For an overview of key questions in social science research about climate change, see Box 5.1.

5.3.1 Internationalization of research

Climate change research is increasingly global in scope, involving international collaborations, utilizing large-scale quantitative techniques and having inputs into policy (Jasanoff and Wynne, 1998). While coordination is at the international level, most funding and operations are carried out at the national level (Ehlers, 1999). Jasanoff and Wynne ask four questions about the context of this knowledge production that, they argue, are best answered by constructivist policy analysis:

- How do scientists and their societies identify and delimit distinct problems related to climate change that are considered amenable to scientific resolution?
- How do scientists come to know particular facts and causal relationships regarding climate change and to persuade others that their knowledge is credible?
- How do conflicts over risk arise, and how are responses to them handled in a world of conflicting and plural political interests?
- How do human societies and their designated policy actors draw upon scientific knowledge to justify collective action on a worldwide scale? (Jasanoff and Wynne, 1998: 5–6)

> ## Box 5.1 Key questions in social science knowledge about climate change
>
> A range of answers to these questions are canvassed in the four-volume study *Human Choice and Climate Change* (Rayner and Malone, 1998a).
>
> - How do scientists choose to study climate change? How do they form a scientific consensus?
> - How do people decide that climate change is worthy of attention?
> - How do people attribute blame for climate change and choose solutions?
> - How do people choose whom to believe about climate change and at what level of risk do they or should they choose to act?
> - How do people and institutions mobilize support for (or against) policy action on climate change?
>
> - What is the relationship between resource management choices and climate change?
> - How do governments establish where climate change stands in relation to other political priorities? How do they choose the climate change issues around which to formulate goals?
> - How are climate change policy instruments chosen?
> - Why and how did the international community choose to address climate change?
> - How do societies select technologies that cause, mitigate or assist adaptation to climate change?
> - How can research on social or collective action be useful to the global climate change debate?
>
> *Source*: Rayner and Malone (1998b: 72)

The internationalization of disciplines, institutions, resources and methodologies poses many challenges (including the deciphering of a bewildering array of acronyms!). One example is the International Geosphere–Biosphere Programme (IGBP), which 'offers an instructive study in the joint formation of an international institution and the framing of global scientific knowledge' (Jasanoff and Wynne, 57). In turn, the IGBP is one of four global environmental change research programmes identified by the International Council of Scientific Unions (Table 5.1). Globalization of knowledge implies conceptualization of the Earth as a system, providing a basis for integration of disparate data. This privileges some disciplines and approaches over others, particularly favouring those which are 'more tractable at global levels of representation', such as remote sensing (p. 49). Thus the present can be given more weight than the past, and physical processes than biological parameters. In general, human dimensions research is more local and regional in scale than scientific research (Ehlers, 1999). While some generalizations about human impacts can be made at the global scale, useful strategies and solutions need more localized expression.

In the context of the IGBP, the coordination of research efforts and the standardization of data sets are encouraged. The latter requires agreements on classification of parameters, for example categories of vegetation cover. The potential loss of local ecological sense (for example, in lumping diverse communities into a savanna biome) in the name of a global overview raises important questions about appropriate scales of analysis.

In an institutional sense the IGBP is divided into six core projects, each operating under a Scientific Steering Committee. Research is classified, under a three-tier system, into 'core research', 'national or regional research' and 'relevant research' (Jasanoff and Wynne, 1998: 58). Researchers are encouraged to acknowledge affiliation with IGBP in publications; this is seen by Jasanoff and Wynne as a conscious strategy of visibility-raising and community-building. While the apparently hierarchical structure could militate against such a community, they argue that extensive

TABLE 5.1 International global environmental change programmes

Programme	Research focus
World Climate Research Programme (WCRP)	The physical climate system, comprising the dynamics of atmosphere, oceans, land surface and ice sheets
International Geosphere – Biosphere Programme (IGBP)	The global interactions between living and non-living processes, that together underpin the habitability and productivity of our planet
International Human Dimensions Programme on Global Environmental Change (IHDP)	The interactions between human society and its environment on a planetary scale
Diversitas	The structure and function of biological diversity, covering plant, animal and microbial life, on land, in fresh water and at sea

Source: Ehlers (1999)

consultation has enhanced participation in the IGBP. For Jasanoff and Wynne, these 'supranational science programs' provide important study sites for the production of scientific knowledge, 'because their authority is still emergent and the processes through which order is being created are less thoroughly naturalized or socialized than in most national programs' (p. 48). As knowledge grows, these processes of knowledge production will become less visible because they appear more natural, unless highlighted through controversy or conflict.

5.3.2 The use of analogy

Only rarely is the use of analogy in climate change research – indeed, in any use of the long-term environmental record – addressed explicitly. Yet its use is frequent; it is one of the most naturalized parts of the knowledge production process. It is useful, then, to examine one example where it has been critically examined (Meyer *et al.*, 1998). Meyer *et al.* examine a number of different types of

analogue; we will look at the most long-term, the use of prehistory. While prehistoric analogues are likely to be less plausible than ones closely connected in time, the examples used are often the most dramatic ones, and they have a particularly strong pull on the public imagination. Using three examples, Meyer *et al.* show that the process of analogical reasoning is problematic, to say the least.

The examples they use are the collapse of civilizations in the Near East and the Aegean around 2200 BC, the collapse of New Kingdom Egypt around 1100 BC, and the Classic Lowland Maya collapse between AD 800 and AD 1000. Although climatic desiccation, Nile failure and climate change respectively have been implicated by a number of researchers, Meyer *et al.* argue that in none of these examples can climate be considered causal. This is not just a matter of establishing clearer spatial or temporal correlations, or of separating anthropogenic and other processes in the palaeoecological or archaeological records, although these aspects are relevant. Rather, to reiterate points made in Chapters 2 and 4, it is crucial to problematize the relationship between different aspects of social and climatic processes:

> Their clearest and most useful lesson is that the significance and consequences of a climate event, once documented, are not direct functions of its physical characteristics. The consequences, rather, vary with the ways in which a society has organized its relations to its resource base, its relations with other societies, and the relations among its members. A climate event is not a sufficient cause for a societal collapse if the failure of the society to deal adequately with it has not been accounted for. (Meyer *et al.*, 1998: 237)

This is not to argue that studying the past is useless or that climate change is trivial – quite the opposite. Rather, it is to emphasize that the usefulness of the past is to help us understand how climate was experienced socially in different contexts.

> The consequences of a shift in climate are not calculable from the physical dimensions of the shift alone, but require attention to the human dimensions through which they are experienced: to the ways in which humans organize their relations to the biophysical environment and to each other. (Meyer *et al.*, 1998: 275)

5.4 Rethinking deforestation

The question of forest loss, particularly in tropical latitudes, is a key one in many debates related to global environmental change. Since land-use pressures are the key driving force in projected change within rainforest and savanna biomes (Table 3.2, p. 44), large-scale forest destruction has become a powerful symbol of the environmental problems that many of us understand to beset the Earth. Measures of forest cover are crucial elements in regional and global climate modelling, for example through attempts to identify albedo changes and estimate anthropogenic CO_2 emissions. Within this context, forest loss in many developing countries is widely portrayed as rapid, recent and a function of population growth. We look here at critiques of these viewpoints from West Africa and India, each undertaken within the framework of a socially oriented political ecology.

The forests at issue in West Africa lie in the forest–savanna mosaic or 'transition' zone along the northern fringe of the forest zone between Sierra Leone and Nigeria. An influential NASA study of anthropogenic deforestation depicts forest loss of between 69 and 96 per cent in the countries of West Africa over the twentieth century (Figure 5.4). In a detailed study of Kissidougou prefecture in Guinea (Fairhead and Leach, 1996) and an overview of West Africa generally (Fairhead and Leach, 1998), Fairhead and Leach challenge this view. Their critique is made both on empirical grounds and via a re-reading of the understandings of forest change held by different actors in the landscape.

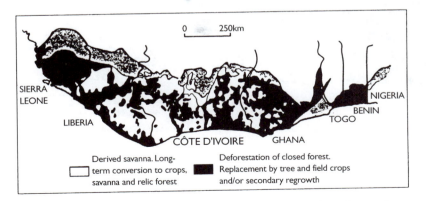

FIGURE 5.4 Schematic map of anthropogenic vegetation changes in West Africa during the twentieth century, as described in an influential NASA study and critiqued by Fairhead and Leach (1998)
Source: Fairhead and Leach (1998: map 1.1) By permission of Taylor & Francis Books Ltd

The empirical grounds involve a detailed study of historical sources, comprising aerial photographs and satellite imagery over different periods, documentary evidence and oral histories. In the case of Kissidougou the photographic data indicate an increase in forest cover in some places and stability in others (e.g. Figure 5.5). The scale of these analyses is such that 'they also show more micro-level dynamics of forest gain and loss. In this they illustrate the dynamic nature of vegetation patterns and the non-unilinear nature of vegetation change' (Fairhead and Leach, 1996: 63). The finding of forest advance is confirmed by the other data sets for Kissidougou. For West Africa generally, Fairhead and Leach argue that twentieth-century deforestation is only about one-third of that suggested by most estimates in broad circulation (Table 5.2). One reason for previous overestimates, they argue, was that the early twentieth-century baseline of most studies was a high point in forest cover, with much recent forest regrowth over lands which had previously been farmed for centuries.

The question of appropriate baselines is just one of the methodological issues that must be dealt with in this type of study. For example, how is forest defined and classified within different sources, and how compatible are they? What is the spatial and spectral resolution of different types of remotely sensed data? There are, however, relatively well-established procedures for resolving those issues and others within an empirical framework. Of more interest here is the broader dilemma in which Fairhead and Leach found themselves:

> On the one hand we are dealing with landscape and its history as representation, but on the other hand we are attempting to reveal its empirical 'reality', facts or events. Very often it is the same sources and data sets which must be used for both. (1996: 15–16)

They do not attempt to escape the ambiguities of this position, nor do they apologize for it. Thus this study provides an example of the type of 'situated knowledge' advocated by Haraway and discussed in Chapter 4.

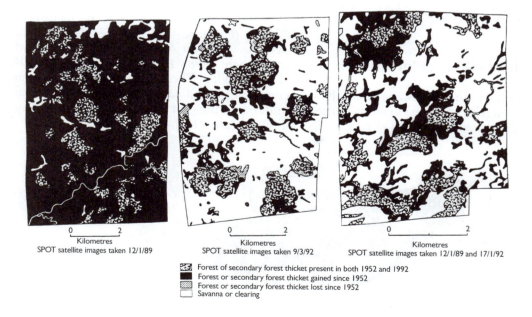

SPOT satellite images taken 12/1/89

SPOT satellite images taken 9/3/92

SPOT satellite images taken 12/1/89 and 17/1/92

Forest of secondary forest thicket present in both 1952 and 1992
Forest or secondary forest thicket gained since 1952
Forest or secondary forest thicket lost since 1952
Savanna or clearing

FIGURE 5.5 Vegetation change in the localities of Toly, Fondambadou and Bamba from 1952 aerial photographs and 1989/92 SPOT satellite images
Source: Fairhead and Leach (1996: figure 2.2, p. 62). By permission of Cambridge University Press

TABLE 5.2 Fairhead and Leach's (1998) suggested revisions to deforestation estimates since 1900

Country	Orthodox estimate of forest area lost	Forest area lost according to World Conservation Monitoring Centre[a]	Suggested revision
Côte d'Ivoire	13	20.2	4.3–5.3
Liberia	4–4.5	5.5	1.3
Ghana	7	12.9	3.9
Benin	0.7	1.6	0
Togo	0	1.7	0
Sierra Leone	0.8–5	6.7	c. 0
Total	25.5–30.2	48.6	9.5–10.5

[a] Figures are in millions of hectares. The World Conservation Monitoring Centre estimates are of forest loss from an undated 'origin' forest cover and are extraordinarily high compared with any other available source. Hence they are presented separately from other orthodox estimates

The Western scientific reading of the Kissidougou landscape as degraded is grounded, it is argued, in a thinking about pristine forest as a baseline of untouched nature against which human impacts can be measured. In contrast, the rural inhabitants of Kissidougou associate forest with a 'settled' rather than a 'natural' state. It is not only human life that has a settled existence in the forest; so too do land spirits and termites. Where the scientists see the savanna as a cultural artefact, and a degraded one at that, it has complex resonances for the local people, associated with vagrant, mobile and impermanent lifestyles (Fairhead and Leach, 1996: 6). The idea, and materiality, of forest expansion as a human artefact is seen in the range of practices used by people in West Africa to foster forests. These include reconversion of savanna into secondary forest by fire exclusion; introduction of preferred trees; encouragement of trees in fallow fields; encouragement of germination of pioneer species and protection of developing forest; and use of termites for soil improvement (Fairhead and Leach, 1996: 289–91).

If the predominant reading of deforestation is so inaccurate, how has it been maintained for so long and become so powerful? This issue is dealt with at length by Fairhead and Leach (1996: ch. 9; 1998: ch. 8), and takes us back to the conditions of knowledge production discussed earlier in this chapter. Particularly powerful was ecological thinking influenced by the Clementsian view of forest as the natural vegetation climax. A convergence of systems thinking in ecology and anthropology contributed to the invocation of 'an image of past society in harmony with a forested environment, resembling the "functional equilibrium" posited in certain cultural ecology studies' (Fairhead and Leach, 1996: 30). The narratives stemming from such views

> depend on – and expose – the field of western imagination and stereotyped images concerning African society … [which] is seen … in terms of a traditional 'functional order' once harmoniously integrated with 'natural' vegetation, and African farming, land and resource-use practices [which] degrade or are at best benign to the original vegetation. (Fairhead and Leach, 1996: 273)

Population growth is read as disrupting this natural order and precipitating land degradation. It follows that degradation can be prevented only by social organization involving interventionist regulation and authority, and denial of local knowledge. The narrative has been reworked in complex ways, it is argued, through colonial forest administrations, scientific institutions, international aid agencies, environmental groups and post-colonial administrations.

In providing this brief rendering of a complex and nuanced set of studies, it is important to emphasize that these authors are not denying that deforestation has occurred at certain times and places, nor that major defor-estation is linked in some places to timber extraction, mining, commercial plantations and farming. The access of the Hyundai Corporation to 300,000 ha of boreal forest in the Bikin river basin in Russia, or of a French logging company to 800,000 ha in Cameroon (Blaikie, 1996), are of different orders of magnitude. It is precisely this attention to scale in understanding the operating mechanisms and processes, as opposed to just the impacts *per se*, that is so important. Nor are Fairhead and Leach arguing for reduced development aid. Rather, they stress that effective policy formulation requires clear problem identifi-cation. Further, in illustrating the links between readings of the landscape and relations of power, they show the range of ways by which local people such as the Kissidougou inhabi-tants, and their knowledge base, have been disempowered. The practical implications of this re-reading of the problem are suggested to include (1) dismantling the dichotomy between preserving 'natural' trees and planting new ones from nurseries; (2) recognition of local concepts of tree tenure; and (3) recognizing people's historical experiences of and relation-ships with currently forested land which is the subject of conservation interest (Fairhead and Leach, 1998: ch. 9). In turn, these and similar studies are contributing to the rethinking of ecological theory, as discussed in Chapter 3.

A useful comparison to this example is provided by Robbins's (1998) analysis of forest discourse in India. Some aspects of the context are the same: understanding of deforestation as driven largely by population pressure; and afforestation programmes promoted in the context of international concern over global environmental change. Robbins is also conscious of the methodological dilemma experienced by Fairhead and Leach: 'how does an analysis of human/environment interaction proceed if it wishes to embrace claims to the historicity of scientific categories, including those of ecological science, while retaining the "observational language" of ecology that might be abandoned in "theoretical anar-chism"?' (Robbins, 1998: 71). Robbins wants to understand, for example, how categories

matter in material circumstances. The specifics of his study are quite the opposite of the West African one. In Rajasthan, forest cover is expanding on paper, as constructed discur-sively by official estimates, while dwindling on the ground.

Official Forest Department definitions of forest in Rajasthan are linked to the idea of enclosed reserves and vegetation dominated by trees. Local experience and categorization is more diverse, but includes open-canopy areas and those subject to multiple use. While dramatic increases in forest cover can be demonstrated in the past few decades, these are mostly plantations of exogenous species. At the same time, traditional community forest-pastures (*orans*) are being lost, to both the plough and Forest Department plantation schemes, partly because of their designation as 'waste' land.

> By introducing exogenous species and limiting the range of local species, the Forest Department's intervention is changing the balance not only of tree cover in village lands but of whole ecological communities. Additionally, these new forest forms integrate poorly with local production systems and create resource gaps for village producers. (Robbins, 1998: 77)

Traditional uses of forest-pastures vary throughout the year, encompassing use of minor tree products, understorey grazing and browsing, and tree browse in the dry season. This usage is patchy in both space and time. 'The establishment of monolithic forests for the sole goal of tree cover production is therefore discordant with the human ecology of the region' (p. 78).

As in the West African case study, there are implications here for discussions of global environmental change. Like Fairhead and Leach, Robbins emphasizes the role of officials, bureaucracies and other institutions in the production of knowledge about the envi-ronment and environmental change. The multiple processes involved in the discursive and material construction of 'forests' affect not just the supposedly empirical issue of esti-mating forest cover. They also 'draw into question the unproblematic linkage of land use to land cover in geographic studies and models of global change' (Robbins, 1998: 82). Others have also argued that an understanding of the human dimension is lacking in conservation and management in the developing world more generally (Saberwal and Kothari, 1996). The most problematic aspect seems to be the assumption that 'all forms of human resource use lead to resource depletion and habitat degradation' (Saberwal and Kothari, 1996: 1329). This has implications both for training of conservation biology professionals and for policy.

5.5 Cultural landscapes as 'discourse materialized'

The examples above relate to discourses about 'forest'. This one is the discourse of 'cultural landscape' itself, and its application in Norwegian and Nordic landscape administration (Jones and Daugstad, 1997). The concept of cultural landscape has a long history of use among Nordic geographers, examples of which were referred to in Chapter 2 (Birks *et al.*, 1988). It has gained currency in different arenas in recent decades with concerns over the effects of urban expansion on the surrounding countryside. By the late 1980s the Norwegian Departments of both the Environment and Agriculture used the concept, and the Directorate for Cultural Heritage established a Landscape Department (Jones and Daugstad, 1997: 268). Increasing availability of funds for consulting and research in the field led to 'cultural landscape' becoming a market opportunity, an academic battleground and a media catchphrase.

Jones and Daugstad analysed '46 documents, published between 1987 and 1993, which dealt with or had relevance for the management of the cultural landscape' (1997: 269). The aim was not to prescribe a correct usage, but to 'identify and discuss problems arising from poorly defined, diverging or contradictory use of concepts in administrative and applied research documents dealing with the cultural landscape in Norway and Norden' (p. 268). The documents included Acts of Parliament and information brochures, as well as policy documents from organizations such as those mentioned above.

The meanings are diverse but also overlap (Table 5.3), and reflect both physical traces of human activity and beliefs associated with landscapes. Some documents were found to distinguish between 'cultural' and 'urban' landscapes, implying that the former was rural. Of particular interest is the sudden appearance in the early 1990s of the concept of 'the cultural landscape of agriculture' in documents from the Department of Agriculture (Jones and Daugstad, 1997: 272–3). This term is used to include the present-day productive agricultural areas, as well as traces of earlier agriculture and remnants of natural vegetation adjoining farmland. 'There are indications that the concept has arisen because "cultural landscape" is seen to be positively value-laden at a time when modern agriculture has met increasing criticism for overproduction and environmentally destructive practices' (p. 273).

This agricultural discourse is one of four identified by Jones and Daugstad, as different groups interpret the meanings of cultural landscape in their own way while competing for influence (Table 5.4). The nature conservation discourse focuses on human-influenced or semi-natural ecosystems considered 'especially valuable' for biodiversity. The cultural heritage sector uses the classical geographical sense of landscapes as modified by human activity, with an emphasis on ancient monuments and historical buildings. There is consciousness here of the interaction between the built and the mental landscapes. In the planning area the influence of architects and landscape architects was dominant. The landscape could be ordered through better siting of constructions, planting of green areas and maintenance of existing vegetation.

Thus 'the cultural landscape provides an arena in which different interest groups struggle to influence the formation of our physical surroundings, exemplified in the conflicts that often arise between the production of economic goods and the production of environmental goods' (Jones and Daugstad, 1997: 280). Jones and Daugstad argue that there are positive possibilities in this dialogue: for example, the focus on processes of environmental change and the human role, and

TABLE 5.3 Summary of meanings given to the term 'cultural landscape' and related concepts in Norwegian and Nordic policy and applied research documents, 1987–93

The Norwegian terms are given in brackets.

1. **Cultural landscape** (*kulturlandskap*)
   (a) Areal category
      (i) cultural landscape v. natural landscape
      (ii) cultural landscape v. urban landscape
   (b) Physical traces of human activity
      (i) chronological process of landscape development leaving visible traces of the past
   (c) Selected elements in the landscape
      (i) threatened cultural monuments or semi-natural vegetation
   (d) Object of management
      (i) natural and cultural heritage
      (ii) agrarian landscapes
   (e) Beliefs and traditions associated with the landscape
      (i) cultural meanings attached to the landscape
      (ii) memories, myths, religions and other beliefs
      (iii) identity
2. **The cultural landscape of agriculture** (*jordbrukets kulturlandskap*)
   Comprises:
   (a) productive farmland
   (b) adjoining natural or semi-natural biotopes
   (c) cultural monuments
   (d) areas earlier used for agriculture
3. **Rural landscape** (*bygdelandskap*)
   (a) Landscape of the countryside (rarely used)
4. **Everyday landscape** (*hverdagslandskap*)
   (a) Ordinary, not especially valuable landscapes, against which valued landscapes can be contrasted
   (b) Ordinary landscapes which require environmental consideration
5. **Traditional and modern landscapes**
   (a) Traditional landscapes = old cultural landscapes with cultural–historical variation and/or botanical diversity, worthy of conservation
   (b) Modern landscapes = production landscapes, productive for agriculture
6. **Landscape prospect or scenery** (*landskapsbilde*)
   (a) Aesthetically pleasing landscape
   (b) Landscape producing visual experience due to variation, cultural heritage, environmental goods or active farming

Source: After Jones and Daugstad (1997: Table 2)

attention to cultural meanings of landscape. However, they see the breadth and multiplicity of the concept as less helpful in the problems of selection in environmental management. For this, much more explicit expression of values and formulation of criteria are required.

Implicit in all the discourses of cultural landscape analysed by Jones and Daugstad is a preservationist tendency: in looking after cultural landscapes we understand ourselves to be looking after aspects of the past. Indeed, it is arguably distress at the ravages of the modern world that provided the context for the cultural landscape concept to emerge around the turn of the twentieth century. As Muir (1998) shows, the early proponents of the historical landscape school, including Carl Sauer and the English writers Jacquetta Hawkes and William Hoskins, were profoundly anti-modernist. The desire to protect past landscapes also extends from the national into the international arena, as we see in the next chapter.

TABLE 5.4 The different discourses of cultural landscape in Norwegian and Nordic policy and applied research documents, 1987–93

Discourse	Emphasis	Rhetoric	Primary values	Other values
Agricultural	Land in production	Agriculture as a producer of amenity values	Economic	Ecological Cultural/ historical
Nature conservation	Semi-natural ecosystems	Maintenance of biodiversity	Ecological	Aesthetic Cultural/ historical
Cultural heritage	Built structures	Significance of physical landscape – meaning	Cultural/ historical	Aesthetic
Planning	Landscape composition	Harmony, tidiness, unity, variety	Aesthetic	Cultural/ historical Ecological

Source: Summarized from Jones and Daugstad (1997)

Part IV

CONTEMPORARY ISSUES AND THE LONG-TERM PERSPECTIVE

PART IV

6

Protecting places

6.0 Chapter summary

This chapter will take the operation of the World Heritage Convention as an influential example of (1) the way the cultural and the natural are demarcated, and (2) the increasing influence, in environmental protection, of international-scale political processes. Changes over the past two decades to the World Heritage definition of, and criteria for inscription as, cultural landscapes are considered. The new category of the 'associative cultural landscape' has emerged in recognition of the intangible dimensions of landscape, and interactions between the physical and the spiritual/symbolic. This process has been influenced by academic debates over culture and nature, by the increasing political voice of indigenous and non-Western peoples, and by the practical difficulties of managing inhabited protected areas. Australia's Uluru–Kata Tjuta National Park, originally nominated under natural heritage criteria, and later successfully renominated as an associative cultural landscape, is used as a case study.

6.1 Mechanisms of protection

How do we ascribe value and significance to places, and what does it mean to try to protect cultural landscapes? This chapter examines these questions in relation to one particular mechanism of ascribing significance and effecting protection, the World Heritage Convention. It forms the backbone of this chapter for two main reasons. First, the interplay between international, national and local forces is an increasingly important one in environmental management (O'Riordan *et al.*, 1998). Second, the attempts of the World Heritage Committee to come to terms with changing conceptions of landscape illustrate the influence of the academic debates outlined in Chapter 4, and show how they play out in management strategies on the ground. These changing conceptions have been influenced not only by academic processes, however, but also by the political voices of non-Western and indigenous peoples in international fora. The chapter discusses the way the World Heritage Convention distinguishes between natural and cultural heritage, and its conceptualizations of cultural landscapes. It examines changes in criteria for inclusion of, and changing descriptions of, cultural properties between 1978 and the late 1990s. These show a shift away from emphases on monuments and buildings, towards considering the physical and social contexts in which structures are found. There is also increasing recognition of the intangible dimensions of landscape, and interactions between the physical and the spiritual/symbolic. At a national and local scale, Australia's Uluru–Kata Tjuta National Park is used as a case study.

Categorizations, of protected and unprotected areas, or of different types of protected areas, need to be recognized and considered as social constructions. This is not to argue that they are any less useful for that. However, it is important to consider how particular constructions have outcomes on the ground. At the most basic level, a particular distinction between the natural and the cultural is maintained in international, national and regional management structures. The World Heritage Committee is advised on natural issues by the International Union for the Conservation of Nature (IUCN), and on cultural issues by the International Committee on Monuments and Sites (ICOMOS). That the whole structure has had to grapple with a broader conceptualization of cultural landscape than just monuments and sites, which in the process challenges unproblematic notions of nature, is the subject of this chapter. Consider, for example, the IUCN categories of protected areas (Table 6.1). The first five categories vary along an axis of 'degree of intervention', with protected areas managed for science or wilderness protection at the 'natural' end and protected areas managed for landscape/ seascape conservation and recreation at the 'artificial' end (Thackway *et al.*, 1996: 20–1). Two issues are notable here. Science is privileged, both as a reason for the highest level of landscape protection, and as an activity focused on natural rather than artificial landscapes. Here science is considered to be an activity outside culture, as we saw in Chapter 5. Second, the sixth category, added later, is the basis for negotiation of indigenous protected areas. Thackway *et al.* note that this category falls about halfway along the conceptual continuum between 'natural' and 'artificial', but it also challenges the whole basis of a protected area classification.

Similar dilemmas pertain in the reflection of the culture–nature distinction in national and regional management structures. 'Cultural heritage' and 'natural heritage' and their equivalents are managed either by different organizations, or different departments within a single organization. Kakadu National Park in Australia provides one example of a structural compartmentalization in management that is discordant with the integration of animals, plants and habitats into the social world of its Aboriginal owners (Allen, 1997). Some have suggested that cultural landscape approaches provide a means of overcoming the divide, and that appears to be a motivation behind increasing World Heritage use of the concept (*see* Box 6.1). Whether it improves things in day-to-day management is explored in the Uluṟu–Kata Tjuṯa case study.

A number of important related issues are touched on in this chapter without being fully explored. Although this discussion focuses on landscapes mostly in public ownership, this is not the only means of protection. Many World Heritage properties are 'parks' or 'reserves' of one sort or another. There are important questions about the implications of such boundary creation for the landscapes outside, an issue that will be considered in Chapter 7. Also, we sidestep here any critical analysis of the notion of 'universal heritage'. That issue will be explored further in Chapter 9.

6.2 World Heritage cultural landscapes

6.2.1 Cultural and natural heritage

The UNESCO Convention Concerning the Protection of the World Cultural and Natural Heritage defines cultural and natural heritage as follows. Cultural heritage includes three categories:

TABLE 6.1 International Union for the Conservation of Nature (IUCN) categories of protected areas

I	Strict Nature Reserve/Wilderness Area: protected area managed mainly for science or wilderness protection
II	National Park: protected area managed mainly for ecosystem protection and recreation
III	Natural Monument: protected area managed mainly for conservation of specific features
IV	Habitat/Species Management Area: protected area managed mainly for conservation through management intervention
V	Protected Landscape/seascape: protected area managed mainly for landscape/seascape conservation and recreation
VI	Managed Resource Protected Area: protected area managed mainly for the sustainable use of natural resources

Source: Phillips (1995: 385)

Box 6.1 World Heritage Convention: paragraphs which deal with categories of cultural landscape

35. With respect to *cultural landscapes* [emphasis in original], the Committee has furthermore adopted the following guidelines concerning their inclusion in the World Heritage List.

36. Cultural landscapes represent the 'combined works of nature and of man' designated in Article 1 of the Convention. They are illustrative of the evolution of human society and settlement over time, under the influence of the physical constraints and/or opportunities presented by their natural environment and of successive social, economic and cultural forces, both external and internal. They should be selected on the basis both of their outstanding universal value and of their representativity in terms of a clearly defined geo-cultural region and also for their capacity to illustrate the essential and distinct cultural elements of such regions.

37. The term 'cultural landscape' embraces a diversity of manifestations of the interaction between humankind and its natural environment.

38. Cultural landscapes often reflect specific techniques of sustainable land-use, considering the characteristics and limits of the natural environment they are established in, and a specific spiritual relation to nature. Protection of cultural landscapes can contribute to modern techniques of sustainable land-use and can maintain or enhance natural values in the landscape. The continued existence of traditional forms of land-use supports biological diversity in many regions of the world. The protection of traditional cultural landscapes is therefore helpful in maintaining biological diversity.

39. Cultural landscapes fall into three main categories, namely:
    (i) The most easily identifiable is the *clearly defined landscape designed and created intentionally by man* [emphasis added]. This embraces garden and parkland landscapes constructed for aesthetic reasons which are often (but not always) associated with religious or other monumental buildings and ensembles.
    (ii) The second category is the *organically evolved landscape* [emphasis added]. This results from an initial social, economic, administrative, and/or religious imperative and has developed its present form by association with and in response to its natural environment. Such landscapes reflect that process of evolution in their form and component features. They fall into two sub-categories:
    - a relict (or fossil) landscape is one in which an evolutionary process came to an end at some time in the past, either abruptly or over a period. Its significant distinguishing features are, however, still visible in material form.
    - a continuing landscape is one which retains an active social role in contemporary society closely associated with the traditional way of life, and in which the evolutionary process is still in progress. At the same time it exhibits significant material evidence of its evolution over time.
    (iii) The final category is the *associative cultural landscape* [emphasis added]. The inclusion of such landscapes on the World Heritage List is justifiable by virtue of the powerful religious, artistic or cultural associations of the natural element rather than material cultural evidence, which may be insignificant or even absent.

(*Source*: www.unesco.org/whc/opgulist.htm# paras 35–39)

- *Monuments*: architectural works, works of monumental sculpture and painting, elements or structures of an archaeological nature, inscriptions, cave dwellings and combinations of features, which are of outstanding universal value from the point of view of history, art or science;
- *Groups of buildings*: groups of separate or connected buildings which, because of their architecture, their homogeneity or their place in the landscape, are of outstanding universal value from the point of view of history, art or science;
- *Sites*: works of man or the combined works of nature and man, and areas including archaeological sites which are of outstanding universal value from the historical, aesthetic, ethnological or anthropological point of view.

The following are considered natural heritage:

- Natural features consisting of physical and biological formations or groups of such formations, which are of outstanding universal value from the aesthetic or scientific point of view;
- Geological and physiographical formations and precisely delineated areas which constitute the habitat of threatened species of animals and plants of outstanding universal value from the point of view of science or conservation;
- Natural sites or precisely delineated natural areas of outstanding universal value from the point of view of science, conservation or natural beauty. (www.unesco.org/whc/world_he.htm Articles 1 and 2)

Cultural properties must meet one of six eligibility criteria (Box 6.2) and natural properties one of four (Box 6.3). Both must pass tests of authenticity and integrity.

Box 6.2   World Heritage Convention: paragraphs which deal with criteria for inclusion of cultural properties in the World Heritage List

24.   A monument, group of buildings or site … which is nominated for inclusion in the World Heritage List will be considered to be of outstanding universal value for the purpose of the Convention when the Committee finds that it meets one or more of the following criteria *and* [emphasis added] the test of authenticity. Each property nominated should therefore:

(a) (i)   represent a masterpiece of human creative genius; or

(a) (ii)   exhibit an important interchange of human values, over a span of time or within a cultural area of the world, on developments in architecture or technology, monumental arts, town-planning or landscape design; or

(a) (iii)   bear a unique or at least exceptional testimony to a cultural tradition or to a civilization which is living or which has disappeared; or

(a) (iv)   be an outstanding example of a type of building or architectural or technological ensemble or landscape which illustrates (a) significant stage(s) in human history; or

(a) (v)   be an outstanding example of a traditional human settlement or land-use which is representative of a culture (or cultures), especially when it has become vulnerable under the impact of irreversible change; or

(a) (vi)   be directly or tangibly associated with events or living traditions, with ideas, or with beliefs, with artistic and literary works of outstanding universal significance (the Committee considers that this criterion should justify inclusion in the List only in exceptional circumstances and in conjunction with other criteria cultural or natural);

and

(b) (i)   meet the test of authenticity in design, material, workmanship or setting and in the case of cultural landscapes their distinctive character and components (the Committee stressed that reconstruction is only acceptable if it is carried out on the basis of complete and detailed documentation on the original and to no extent on conjecture).

(b) (ii) have adequate legal and/or contractual and/or traditional protection and management mechanisms to ensure the conservation of the nominated cultural properties or cultural landscapes. The existence of protective legislation at the national, provincial or municipal level and/or a well-established contractual or traditional protection as well as of adequate management and/or planning control mechanisms is therefore essential and, as is clearly indicated in the following paragraph, must be stated clearly on the nomination form.

Assurances of the effective implementation of these laws and/or contractual and/or traditional protection as well as of these management mechanisms are also expected. Furthermore, in order to preserve the integrity of cultural sites, particularly those open to large numbers of visitors, the State Party concerned should be able to provide evidence of suitable administrative arrangements to cover the management of the property, its conservation and its accessibility to the public.

(*Source*: www.unesco.org/whc/opgulist.htm# para 24

Box 6.3   World Heritage Convention: paragraphs which deal with criteria for inclusion of natural properties in the World Heritage List

44. A natural heritage property – as defined above – which is submitted for inclusion in the World Heritage List will be considered to be of outstanding universal value for the purposes of the Convention when the Committee finds that it meets one or more of the following criteria and fulfils the conditions of integrity set out below. Sites nominated should therefore:

(a) (i) be outstanding examples representing major stages of Earth's history, including the record of life, significant on-going geological processes in the development of land forms, or significant geomorphic or physiographic features; or

(a) (ii) be outstanding examples representing significant on-going ecological and biological processes in the evolution and development of terrestrial, fresh water, coastal and marine ecosystems and communities of plants and animals; or

(a) (iii) contain superlative natural phenomena or areas of exceptional natural beauty and aesthetic importance; or

(a) (iv) contain the most important and significant natural habitats for in-situ conservation of biological diversity, including those containing threatened species of outstanding universal value from the point of view of science or conservation;

and

(b) also *fulfil the following conditions of integrity* [emphasis in original]:

(b) (i) The sites described in 44(a)(i) should contain all or most of the key interrelated and interdependent elements in their natural relationships; for example, an 'ice age' area should include the snow field, the glacier itself and samples of cutting patterns, deposition and colonization (e.g. striations, moraines, pioneer stages of plant succession, etc.); in the case of volcanoes, the magmatic series should be complete and all or most of the varieties of effusive rocks and types of eruptions be represented.

(b)(ii) The sites described in 44(a)(ii) should have sufficient size and contain the necessary elements to demonstrate the key aspects of processes that are essential for the long-term conservation of the ecosystems and the biological diversity they contain; for example, an area of tropical rain-forest should include a certain amount of variation in elevation above sea-level, changes in topography and soil types, patch systems and naturally regenerating patches; similarly a coral reef should include, for example, seagrass, mangrove or other adjacent ecosystems that regulate nutrient and sediment inputs into the reef.

(b)(iii) The sites described in 44(a)(iii) should be of outstanding aesthetic value and include areas that are essential for maintaining the beauty of the site; for example, a site whose scenic values depend on a waterfall, should include adjacent catchment and downstream areas that are integrally linked to the maintenance of the aesthetic qualities of the site.

(b)(iv) The sites described in paragraph 44(a)(iv) should contain habitats for maintaining the most diverse fauna and flora characteristic of the biogeographic province and ecosystems under consideration; for example, a tropical savannah should include a complete assemblage of co-evolved herbivores and plants; an island ecosystem should include habitats for maintaining endemic biota; a site containing wide-ranging species should be large enough to include the most critical habitats essential to ensure the survival of viable populations of those species; for an area containing migratory species, seasonal breeding and nesting sites, and migratory routes, wherever they are located, should be adequately protected; international conventions, e.g. the Convention of Wetlands of International Importance Especially as Waterfowl Habitat (Ramsar Convention), for ensuring the protection of habitats of migratory species of waterfowl, and other multi- and bilateral agreements could provide this assurance.

(b)(v) The sites described in paragraph 44(a) should have a management plan. When a site does not have a management plan at the time when it is nominated for the consideration of the World Heritage Committee, the State Party concerned should indicate when such a plan will become available and how it proposes to mobilize the resources required for the preparation and implementation of the plan. The State Party should also provide other document(s) (e.g. operational plans) which will guide the management of the site until such time when a management plan is finalized.

(b)(vi) A site described in paragraph 44(a) should have adequate long-term legislative, regulatory, institutional or traditional protection. The boundaries of that site should reflect the spatial requirements of habitats, species, processes or phenomena that provide the basis for its nomination for inscription on the World Heritage List. The boundaries should include sufficient areas immediately adjacent to the area of outstanding universal value in order to protect the site's heritage values from direct effects of human encroachment and impacts of resource use outside of the nominated area. The boundaries of the nominated site may coincide with one or more existing or proposed protected areas, such as national parks or biosphere reserves. While an existing or proposed protected area may contain several management zones, only some of those zones may satisfy criteria described in paragraph 44(a); other zones, although they may not meet the criteria set out in paragraph 44(a), may be essential for the management to ensure the integrity of the nominated site; for example, in the case of a biosphere reserve, only the core zone may meet the criteria and the conditions of integrity, although other zones, i.e. buffer and transitional zones, would be important for the conservation of the biosphere reserve in its totality.

Sites described in paragraph 44(a) should be the most important sites for the conservation of biological diversity.

Biological diversity, according to the new global Convention on Biological Diversity, means the variability among living organisms in terrestrial, marine and other aquatic ecosystems and the ecological complexes of which they are part and includes diversity within species, between species and of ecosystems. Only those sites which are the most biologically diverse are likely to meet criterion (iv) of paragraph 44(a).

(*Source*: www.unesco.org/whc/opgulist.htm#para 44)

All the three types of cultural properties would constitute cultural landscapes within the understanding of this book. Indeed, the fact that natural areas and sites are valued, protected and managed in terms of something called 'world heritage' makes them also cultural landscapes. They are brought into the human realm by being interpreted and classified in a certain way. But, as was explored in Chapter 4, this problematization of the 'natural' is a relatively recent academic trend, one to which the World Heritage Committee is itself responding by its increased recognition of associative cultural landscapes without necessary material alteration. For example, Yellowstone and Yosemite National Parks in the USA were inscribed as World Heritage natural properties in 1978 and 1984 respectively. These definitive wilderness areas have been used as prime examples in recent discussions that show the multiple cultural bases of such landscapes. These include seeing them:

- as early examples of the National Park concept which would become influential worldwide;
- as icons of the sublime tradition in white North American thinking about nature;
- as sites of Native American occupation and activity, including manipulation through the use of fire (e.g. Cronon, 1996a; Olwig, 1996; Spirn, 1996).

If these properties were nominated today, it would most likely be on the basis of cultural as well as natural criteria (N.J. Mitchell, 1995). The property discussed in detail in this chapter, Uluru–Kata Tjuta National Park, was originally inscribed on the basis of natural criteria in 1987, and reinscribed in terms of cultural criteria in 1994 in recognition of it as an Anangu (the name of the local Aboriginal people) landscape.

There are important issues here that relate to the way this key mechanism of international heritage protection distinguishes between culture and nature, and promulgates a notion of universal value. Following discussions in Chapters 4 and 5, we should also pay attention to the way in which these international negotiations construct and influence certain understandings of and knowledge about landscape. Although these issues are not the focus of this chapter, it is important to acknowledge them before moving on. Take as a further example the Australian Fossil Mammal Sites (Riversleigh/Naracoorte), inscribed in 1994 under natural criteria (a) (i) and (ii) (*see* Box 6.3 for details of these criteria). In the north and south respectively of eastern Australia, they are considered 'among the world's ten greatest fossil sites. They superbly illustrate the stages of evolution of Australia's unique fauna' (www.unesco.org/whc/sites). Without denying the significance of the sites or their fossil record, it is important to remember the profoundly cultural processes that rendered them significant. The classification of these sites as 'natural' makes the human labour of science invisible. Yet the public meaning of these sites has been enhanced because of exposure to that very labour: the 'detective work' of discoveries, the romance and tedium of remote-area fieldwork, and the 'painstaking' laboratory processing (Archer *et al.*, 1994). Indeed, it is unlikely that there would have been pressure for the World Heritage nomination without the construction of Riversleigh within the public imagination. As we saw in Chapter 5, science needs to be explicitly recognized as part of these cultural processes.

However, while being mindful of the profound cultural basis of the designation of some sites as World Heritage 'nature', we are going to focus in this case study on the specific properties designated as cultural landscapes, a concept adopted by the World Heritage Committee only in 1992 (UNESCO, 1995). Cultural landscapes are ones which represent the 'combined works of nature and of man' as designated under Article 1 of the convention, above. The inclusion of this category was an attempt to go beyond seeing cultural heritage purely in terms of monuments, and recognized that the cultural and natural elements may be inseparable. They fall into three main categories, as defined under paragraph 39 of the Operational Guidelines (Box 6.1). The first and most easily identifiable is the clearly defined and intentionally created landscape, such as gardens and parkland. The second is the organically evolved landscape, which can be either relict (fossil) or continuing. An example of a fossil landscape is one with very visible prehistoric sites, such as tumuli linked to funeral ceremonies in the Sahara. Continuing landscapes include the irrigated rice terraces of Luzon in the northern Philippines. These are a very obvious landscape of physical transformation, with the terraces supported by dry-stone walls on steep mountainsides 700–1500 m above sea level. There are also conceptual expressions here: 'each of the clusters of terraces has a buffer ring of private forests (*muyong*), a hamlet and a sacred grove'

(Titchen, 1996: 34). Villages are focused on a ritual rice field, always the first to be planted and harvested.

The third category is the associative cultural landscape, whereby powerful religious or cultural associations with nature may have little or no material evidence. An example is the sacred wood in the Harare area of Zimbabwe, which has been preserved from farming pressure by religious taboos against chopping it down (UNESCO, 1999). The addition of the associative cultural landscape acknowledges the link between the physical and spiritual aspects of landscape. It was seen to be particularly relevant to the cultural practices of indigenous peoples, and to the Asia-Pacific region with its diversity of living traditions in relation to land and water (UNESCO, 1995). The intellectual context in which such concepts could gain political ground was influenced by discussions on landscape occurring within a number of academic disciplines, such as those discussed in Chapter 4. The first two properties to be included on the World Heritage list for their associative cultural values were Tongariro National Park in New Zealand and Uluru–Kata Tjuta National Park in Australia (discussed in more detail in section 6.3). Both properties had already been listed for their natural values. This highlights the close correlation between traditional custodianship by indigenous people and the maintenance of qualities described as 'natural heritage', and begs the question of the extent to which environments valued for their 'naturalness' are in fact artefacts of indigenous activity over millennia. Further, these shifts in thinking are not without their challenges on the ground; for example, 'management problems may arise if traditional land-use practices are seen to conflict with other nature conservation strategies' (UNESCO, 1995: 11) (*see* Chapter 8).

6.2.2 Changes in cultural landscapes, 1978–1997

In Chapter 4 we considered how thinking about landscapes has changed over the past two decades or so. During this period World Heritage thinking about cultural landscapes has also changed in response to both academic and political processes. In turn, the recognition of a broader range of cultural landscapes will influence public thinking, as can be seen in the example of Uluru–Kata Tjuta (section 6.3).

The World Heritage Committee has recognized cultural landscapes since 1992, with attention before that time focused on the two other categories of monuments and buildings. More subtle changes in thinking and practice are reflected in changes since 1978 to the criteria by which all cultural properties have been assessed (compare Box 6.2 with Box 6.4). There is a shift from unidirectional cultural influence to 'interchange' (criterion (ii)). Emphasis on rarity or antiquity for its own sake has been removed, and the possibility of living traditions included (criterion (iii)). Structures or buildings are now considered more within their landscape context, as part of an 'ensemble', and also within historical context (criterion (iv)). An emphasis on urban structures has changed to the broader 'human settlement and land use' (criterion (v)). Within the criterion related to ideas and beliefs (criterion (vi)), emphasis on historical events and persons has been replaced with consideration of outstanding universal significance. Overall, then, the criteria have broadened to offer the possibility of considering more

dynamic and symbiotic relationships between people and their environment, and to avoid a linear view of history (Rossler, 1995). This does not mean that monuments and built structures are any less important, but they are likely to be considered more in broader context, rather than just, say, for architectural or artistic merit.

These changes in thinking can also be illustrated by examining changes in the types of cultural properties included on the World Heritage list over time. It is important to remember that the process of nomination and inclusion is itself a complex cultural process and a product of many factors, including domestic political ones in the nominating state. We need to reiterate that the changes reflected here are changes in how people think about and value landscapes. They do not of themselves represent any absolute change in the abundance of certain types of landscapes. (There may, however, be consequent changes on the ground, for example if the environs of a significant building are brought into management as part of a more contextual approach.) Some of the processes are apparently contradictory. The globalizing structure of the UN is mirrored in the rhetoric of 'universal significance'. On the other hand, the move towards the 'associative cultural landscape' is a move towards recognizing the

---

Box 6.4 World Heritage cultural heritage criteria, 1978

(i)   Represent a unique artistic or aesthetic achievement, a masterpiece of the creative genius.

(ii) Have exerted considerable influence, over a span of time or within a cultural area of the world, on development in architecture, monumental sculpture, garden and landscape design, related arts, town-planning or human settlements

(iii) Be unique, extremely rare, or of great antiquity.

(iv) Be among the most characteristic examples of a type of structure, the type representing an important cultural, social, artistic, scientific, technological or industrial development.

(v)   Be a characteristic example of a significant style of architecture, method of construction or form of town-planning or traditional human settlement that is fragile by nature or has become vulnerable under the impact of irreversible socio-cultural or economic change.

(vi) Be most importantly associated with ideas or beliefs, with events or with persons, of outstanding historical importance or significance.

Source: www.unesco.org/whc/c-criteria-changes.htm. Various individual criteria were changed incrementally in 1980, 1984, 1994 and 1997.

diversity of environmental interactions expressed at the local scale.

Early cultural properties inscribed on the World Heritage List (for detailed descriptions of sites, and all references in this section, *see* www.unesco.org/whc/sites) include buildings and monuments with important art associated:

- Boyana Church, Sofia (Bulgaria, inscribed 1979), 'which is covered with frescoes, painted in 1259, making it one of the most important collections of medieval painting';
- the church and Dominican convent of Santa Maria delle Grazie with *The Last Supper* by Leonardo da Vinci (Italy, inscribed 1980).

Even where a total landscape is being inscribed, there is a tendency to describe and justify it with reference to architecture and art. For example, Venice and its Lagoon (Italy, inscribed 1987) 'is, as a whole, an extraordinary architectural masterpiece in which even the smallest of its buildings contains the works of some of the world's greatest artists such as Giorgione, Titian, Tintoretto, Vernonese and others'.

Indigenous or archaeological heritage is recognized in early inscriptions, where it can be related to built structures:

- the village of Ninstints, Anthony Island (Canada, inscribed 1981), with houses and 32 totem and mortuary poles;
- Mesa Verde (USA, inscribed 1978), a concentration of Anasazi Indian dwellings built from the sixth to the twelfth centuries;
- Stonehenge, Avebury and associated sites (United Kingdom, inscribed 1986);
- Maya site of Copan (Honduras, inscribed 1980);
- Archaeological ruins at Moenjodaro (Pakistan, inscribed 1980).

Where criterion (vi), relating to ideas and beliefs, was invoked, it was likely to be in relation to a religious building or monument:

- Aachen Cathedral (Germany, inscribed 1978);
- the Pyramid fields from Giza to Dahshur (Egypt, inscribed 1979);
- Durham Castle and Cathedral (United Kingdom, inscribed 1986);
- Canterbury Cathedral, Saint Augustine's Abbey, and Saint Martin's Church (United Kingdom, inscribed 1988);

or to a place associated with one or more significant historical events:

- Independence Hall (USA, inscribed 1979);
- Palace and Park of Versailles (France, inscribed 1979);
- L'Anse aux Meadows National Historic Park (Canada, inscribed 1978), site of the first historic traces of a European presence in the Americas – an eleventh-century Viking settlement.

Among more recent nominations, there is a continuation of interest in significant archaeological structures (e.g. archaeological site of Troy, Turkey, inscribed 1998) and outstanding historic monuments (e.g. historic monuments of ancient Nara, Japan, 1998). There is also a noticeable tendency for 'ensembles' of buildings or designed landscapes to be more explicitly considered in their spatial and social contexts (e.g. the Millenary Benedictine Abbey of Pannonhalma and its natural environment, Hungary, inscribed 1996; the Lednice–Valtice cultural landscape, Czech Republic, inscribed 1996). Thus religious ensembles are described more in contextual than architectural terms:

- *Ouadi Qadisha (the Holy Valley) and the forest of the Cedars of God (Horsh Arz el-Rab) (Lebanon, inscribed 1998)*: 'The Qadisha Valley is one of the most important early Christian monastic settlements in the world, and its monasteries, many of great age, are positioned dramatically in a rugged landscape. Nearby are the remains of the great forest of the Cedars of Lebanon, which were highly prized for the construction of great religious buildings in Antiquity'.
- *Flemish Béguinages (Belgium, inscribed 1998)*: 'The Beguines were women who entered into a life dedicated to God without retiring from the world. In the 13th century they founded the béguinages, enclosed communities designed to meet their spiritual and material needs. The Flemish

Béguinages are architectural ensembles composed of houses, churches, ancillary buildings and green spaces organized according to a spatial conception of urban or rural origin, and are built in styles specific to the Flemish cultural region. They bear extraordinary witness to the tradition of the Beguines that developed in north-western Europe in the Middle Ages.'

There is increasing recognition of the value of prehistoric artistic expression and marking of the landscape. These tend to be described in universalist rather than local terms:

- *Prehistoric rock art sites in the Côa Valley (Portugal, inscribed 1998)*: 'an outstanding example of the sudden flowering of creative genius at the dawn of human cultural development', in the period from 22,000 to 10,000 BC.
- *Rock art of the Mediterranean Basin on the Iberian Peninsula (Spain, inscribed 1998)*: provides 'an exceptional picture of human life in a seminal period of human cultural evolution'.

Where living peoples express a connection to the land and to ancestors over periods at least as long, as in Uluru–Kata Tjuta, for example, the description is more in terms of local attachments than universal ancestors, as we shall see below.

A number of important properties continue to be inscribed for meeting a combination of natural and cultural criteria. These include recognition of traditional interactions with landscape, including in Europe:

- *The Laponian Area (Sweden, inscribed 1996)*, home of the Saami people. 'Every summer, the Saami lead their immense herds of reindeer towards the mountains through a natural landscape hitherto preserved, but now threatened by the advent of motor vehicles. Historic and on-going geological processes can be seen in the glacial moraines and changing water courses.' (*see* Chapter 9 for more detailed discussion of Sámi issues.)
- *Pyrénées – Mount Perdu (France/Spain, inscribed 1997)*: 'The site … includes two of Europe's largest and deepest canyons on

the Spanish side and three major cirque walls on the more abrupt northern slopes with France, classic presentations of these geologic landforms. But the site is also a pastoral landscape reflecting an agricultural way of life that was once widespread in the upland regions of Europe. It has survived unchanged into the 20th century only in this part of the Pyrénées, providing exceptional insights into past European society through its landscape of villages, farms, fields, upland pastures, and mountain roads.'

In keeping with the efforts of the World Heritage committee to inscribe more properties from the Asia-Pacific region, nominations invoking both cultural and natural criteria have come from China, e.g. Mt Emei and Leshan Giant Buddha (1996). Emei is the site of a number of temples, and the largest Buddha in the world, and is 'also notable for its very diverse vegetation, ranging from sub-tropical to subalpine forests. Some of the trees are more than a thousand years old.' In inscribing this property the World Heritage committee underlined the importance of the link between the tangible and the intangible, the natural and the cultural (UNESCO, 1997: B.1). Chinese properties inscribed under purely cultural criteria also refer to harmonious interactions between humankind and nature:

- *Lushan National Park (inscribed 1996)*: 'Mount Lushan, in Jiangxi, is one of the spiritual centres of Chinese civilization. Buddhist and Taoist temples, along with landmarks of Confucianism, where the most eminent masters taught, blend well into a strikingly beautiful landscape which has inspired countless artists who developed the aesthetic approach to nature found in Chinese culture.'
- *Summer Palace, an imperial garden in Beijing (inscribed 1998)* 'is a masterpiece of Chinese landscape garden design, integrating the natural landscape of hills and open water with manmade features such as pavilions, halls, palaces, temples and bridges into a harmonious and aesthetically exceptional whole'.

6.3 Uluru–Kata Tjuta National Park

To Euro-Australian eyes, Uluru (also known as Ayers Rock) is a sandstone monolith some 9.4 km in circumference and rising about 340 m above the surrounding plain. To Australians it and Kata Tjuta (the Olgas) represent the centre, the desert, the outback. To them and non-Australians alike, 'the rock' is one of the iconic images of Australia. The outstanding natural scenery of Uluru–Kata Tjuta National Park was a key justification for its 1987 nomination to and inscription on the World Heritage List as a natural property (Box 6.5). The nomination also referred to 'outstanding examples of tectonic geochemical and geomorphic processes' and the surrounding arid land ecosystems (Australia, 1999: Appendix A). The then natural heritage criterion (criterion (iii)) recognized an 'exceptional combination of natural and cultural elements', and it was noted that the 'overlay of Aboriginal occupation adds a fascinating cultural aspect to the site' (Australia, 1994: 24).

For Anangu, whose ancestors have lived in the area for at least 30,000 years, Uluru-Kata Tjuta is home – a cultural rather than a natural landscape. Accordingly it was renominated under cultural criteria in 1994 (Box 6.5). The 'rock' is not a single place, but a complex of individually named places (Figure 6.1). Today's landscape is the outcome of millennia of management using traditional methods governed by the *Tjukurpa* (the law),

> which describes how the land was given form as a result of the travels of ancestral beings, and indeed defines all aspects of the relationship between people and the land and between people themselves. …

The *Tjukurpa* is a complex concept which combines ancestral time and present time. It is founded upon a time when heroic beings who combined the attributes of humans and animals camped and travelled across the landscape. As they did so they shaped the features of the land. Their bodies, artefacts and actions became places imbued with their presence. (Australia, 1994: 8, 21)

Anangu interact with the landscape in a variety of ways that scientists tend to think of in ecological terms. They apply a mosaic burning regime; they hunt and gather plants and animals; they use mobility and reciprocal rights with neighbouring groups as a buffer against the harsh and highly variable climate. For Anangu, however, these rights and responsibilities are prescribed and governed by *Tjukurpa*, which provides the basis for all social

---

Box 6.5   Nomination of Uluru–Kata Tjuta National Park, Australia, for inscription on the World Heritage List as a cultural landscape, 1994

In accordance with the following operational guidelines (*see* Box 6.1 for paragraph references):

- As a cultural landscape representing 'the combined works of nature and of man' (paragraph 36), manifesting the 'interaction between humankind and its natural environment' (paragraph 37);
- As a cultural landscape reflecting 'specific techniques of sustainable land-use' and 'a specific spiritual relation to nature' that can contribute to modern techniques of sustainable land-use' and 'support biological diversity' (paragraph 38);
- As an associative cultural landscape having 'powerful religious, artistic' and 'cultural associations of the natural element' (paragraph 39(iii)).

    Uluru–Kata Tjuta is nominated under the following criteria:

24(a)(v)   an outstanding example of a traditional human settlement and land-use which is representative of a culture (or cultures), especially when it has become vulnerable under the impact of irreversible change.

24(a)(vi)   directly and tangibly associated with events or living traditions, with ideas, or with beliefs … of outstanding universal significance.

Source: Summarized from Australia (1994: ch. 5)

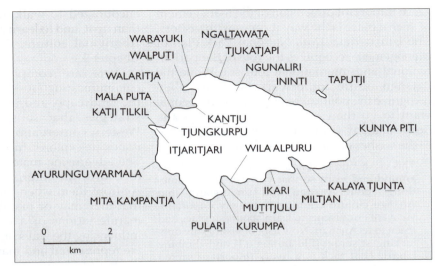

Figure 6.1 Some Anangu place names at Uluru
Source: Uluru-Kata Tjuta National Park Draft Plan of Management (figure 4)

processes (Australia, 1994). *Tjukurpa* thus provides the moral – and, under the joint management agreement between Anangu and Parks Australia, the legal – basis for the management of Uluru–Kata Tjuta National Park, whose board of management has an Aboriginal majority (Australia, 1999).

What are the implications of managing this place as a cultural rather than a natural landscape, of privileging *Tjukurpa*? There is much common ground with Western scientific principles of ecological management, and it would be wrong to overstate the differences. However, a few examples illustrate the significance of those differences (for more detailed discussion, *see* Australia, 1999).

In contrast to many Western approaches which would see human activity as a threat to biodiversity, maintenance – or, where they have been interrupted by colonization, reinstitution – of Anangu practices enhances biodiversity. For example, maintenance of the vegetation mosaic in the spinifex grasslands by regular burning has been shown to be essential for the survival of a number of endemic species, including mulgara (*Dasycercus cristicauda*) (Australia, 1994). Maintenance of a network of wells and waterholes across the country contributed to wildlife refugia, and the expertise of senior Anangu women is being used to re-establish this maintenance.

Joint research by Western scientists and Anangu experts has resulted in new insights, and is particularly important to faunal management. Anangu recognize four principal ecological zones; this classification parallels but is not identical to a Western scientific classification (Table 6.2).

Table 6.2 Comparison of ecological zones recognized by Anangu and Western land systems within Uluru–Kata Tjuta National Park

Anangu zones	Western land systems
Apu or *puli* – rock faces and vegetated hill slopes;	*Gillen* – the two large outcrops (Uluru–Kata Tjuta), their gorges, gullies and creeklines, and associated fans and alluvium;
Puti – woodlands, particularly the mulga flats between sandhills;	*Karee* – gently sloping plains fringing the fans and alluvium of the *Gillen*;
Tali and *pila* – sand dunes and sand plains;	*Simpson* – sand plains and dune fields
Karu – creek beds	

Source: Australia (1994: 32; 1999: 65)

An important dimension of the management of such a place is the way it copes with visitors. The Uluru–Kata Tjuta National Park Plan of Management recognizes the role of the park in national and international tourism. While the Anangu desire for privacy is strong, they recognize the political reality that many people want to go there. A central dilemma is the desire of many visitors to climb Uluru, contrary to the wishes of the traditional owners.

> This is Anangu land and we welcome you. Look around and learn so that you can know something about Anangu and understand that Anangu culture is strong and really important. We want our visitors to learn about our place and listen to us Anangu. Now a lot of visitors are only looking at sunset [Illustration 6.1] and climbing Uluru. That rock is a really important sacred thing. You shouldn't climb it! Climbing is not a proper part of this place. (Australia, 1999: 109)

This quotation from senior owner Tony Tjamiwa emphasizes the spiritual disrespect for *Tjukurpa* that climbing the rock shows. There is also a safety aspect, since the *Tjukurpa* gives Anangu the responsibility, or 'duty of care', for visitors to their country.

Tjamiwa also presents the broader context in which Anangu seek to welcome visitors: to have Uluru–Kata Tjuta understood as first and foremost an Anangu place. A key means of education in this respect is the cultural centre at the park, as well as increasing Anangu involvement as tour guides, educators and rangers (Illustration 6.2). Thus visitors are encouraged to walk around the rock rather than up it, and to learn about various aspects of Aboriginal culture. (For further details, *see* Chapter 9.)

There are complex issues involved in communicating cross-cultural ways of seeing the landscape, and in devising management strategies that satisfy both *Tjukurpa* and Western conservation values. Some of those issues are explored further in Chapters 8 and 9. The advertising material that attracts tourists to Uluru, the interpretive signage that will confront them when they arrive, the fire scars that they may or may not notice are all local manifestations of a nested set of processes influencing the ways highly valued landscapes are represented and managed. In this example the World Heritage inscription of Uluru–Kata Tjuta as an associative cultural landscape provides an international mechanism for the recognition of Anangu aspirations. In turn, recognition of this conceptualization as the central one influences outcomes on the ground.

In considering tensions in the management of Uluru–Kata Tjuta, of which the issue of climbing the rock provides just one example, it is important to remember that the Aboriginal peoples' land claim to the area was conditional on their leasing it back to the Commonwealth of Australia as a national park. Nominal Aboriginal power is constrained by political realities. If there are tensions and dissatisfactions in the most widely cited model examples of joint management, Uluru–Kata Tjuta and

ILLUSTRATION 6.1 Tourists watch the sunset at Uluru (by permission of Richard Fullagar)

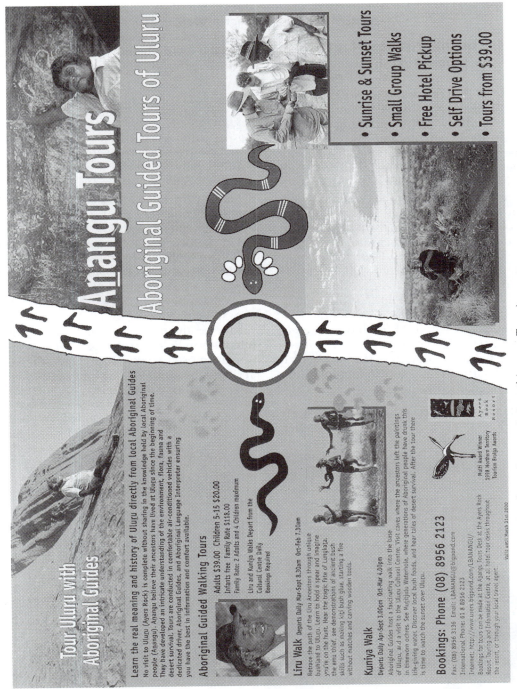

ILLUSTRATION 6.2 Anangu tour brochure (by permission of Anangu Tours)

Tour Uluru with Aboriginal Guides

Anangu Tours

Aboriginal Guided Tours of Uluru

Learn the real meaning and history of Uluru directly from local Aboriginal Guides

No visit to Uluru (Ayers Rock) is complete without sharing in the knowledge held by local Aboriginal people (Anangu). Anangu believe their ancestors have lived at Uluru since the beginning of time. They have developed an intricate understanding of the environment; flora, fauna and desert survival. Tours are conducted in comfortable air-conditioned vehicles with a dedicated driver, Aboriginal Guides, and Aboriginal Language Interpreter ensuring you have the best in information and comfort available.

Aboriginal Guided Walking Tours

Adults $39.00 Children 5-15 $20.00
Infants free Family Rate $118.00
Family Rate: 2 Adults and 4 Children maximum

Liru and Kuniya Walks Depart from the
Cultural Centre Daily
Bookings Required

Liru Walk Departs Daily Mar-Sept 8.30am Oct-Feb 7.30am

Retrace the path of the Liru Ancestors through unique bushland to Uluru. Learn to hold a spear and imagine you're on the hunt. Hear the tragic fate of Lungkata, the emu thief; see demonstrations of ancient bush skills such as making kiti bush glue, starting a fire without matches and carving wooden tools.

Kuniya Walk

Departs Daily Apr-Sept 3.00pm Oct-Mar 4.00pm

Aboriginal Guides host a fascinating walk into the base of Uluru, and a visit to the Uluru Cultural Centre. Visit caves where the ancestors left the paintings in timeworn ochres. See Mutitjulu Waterhole, where generations of Aboriginal people have drunk this life-giving water. Discover local bush foods, and hear tales of desert survival. After the tour there is time to watch the sunset over Uluru.

Bookings: Phone (08) 8956 2123

Fax: (08) 8956 3136 Email: LBANANGU@bigpond.com
International: Phone: 61 8 8956 2123
Internet: http://www.users.bigpond.com/LBANANGU/
Bookings for tours can be made at the Anangu Tours Desk at the Ayers Rock Resort Touring and Information Centre, at all hotel tour desks throughout the resort, or through your local travel agent.

Multi Award Winner
1998 Northern Territory
Tourism Brolga Awards

Valid until March 31st 2000

Ayers Rock Resort

- Sunrise & Sunset Tours
- Small Group Walks
- Free Hotel Pickup
- Self Drive Options
- Tours from $39.00

Kakadu National Parks, these are even greater in situations where Aboriginal people have less power over land (Smyth and Sutherland, 1996).

6.4 The inheritance of future generations

At one level the connection of World Heritage cultural landscapes with the past is so obvious as to be hardly worth mentioning. Whether explicitly designed, organically evolved or associative, these are landscapes that attest to long periods of human interaction. Their significance accrues partly because of the passage of time. Even in those landscapes where tradition is seen as a living thing, it is also valued in terms of its continuity with the past. The documentation for the World Heritage listing of Uluru–Kata Tjuta National Park, for example, makes frequent reference to the fact that Anangu ancestors have lived in the area for more than 30,000 years, and that the contemporary Anangu social structure and adaptations are thought to have emerged within the past 5000 years.

Yet the notion of protecting the past leads to the use of particular imagery in terms of future action. The counterposing of long-accumulated past time against the 'rapidly developing world' accentuates the possibility of future 'threat' and 'damage'. World Heritage (or any other protective mechanism for that matter) then presents itself as a bulwark against the otherwise overwhelming tide of rapid and destructive change. For example,

> With 582 cultural and natural sites already protected worldwide, the World Heritage Centre is working to make sure that future generations can inherit the treasures of the past. And yet, most sites face a variety of threats, particularly in today's environment. The preservation of this common heritage concerns us all. (www.unesco.org/whc)

It is not my intention to underrate the threats to important cultural and natural sites – quite the opposite. Rather it is to show how the preservationist rhetoric unduly constrains discussions of the problem. The subtext, that the ideal would somehow be to freeze time, does not help us think about managing for change. This forms the subject of Chapter 7. Nor does there seem to be much conceptual space among the threats for the possibility of positive human influences in living landscapes. Yet an ongoing human presence is integral to all World Heritage properties, cultural and natural. These dilemmas are addressed in Chapter 8.

7

Restored, (p)reserved and created landscapes

7.0 Chapter summary

How do we manage landscapes in an environment of change? Three key terms of human intervention in environmental and cultural heritage management – *restoration, preservation* and *creation* – each contain an implied temporal direction. The issues involved in going backwards, maintaining the present and shaping the future are discussed, using examples from both biotic and built landscapes. The value of the long-term record is not that it specifies previous conditions which should be restored, but that it can illustrate the processes and mechanisms of change, and the likely location of irreversible thresholds. The possibilities and context of restoration in recently settled New Zealand and long-occupied parts of the Old World are compared. The cultural heritage management principle that the restoration process must explicitly draw attention to itself, and label itself as such, provides a good model for ecological restoration. Landscapes of preservation, as exemplified by national parks, need to incorporate a capacity for change, whether at the level of wildlife migration patterns, political variables or climatically induced vegetation shifts. In all these examples, clarification of management goals is crucial to effective intervention.

7.1 Managing for change

The central issue of this chapter is the management of landscapes in a context of environmental change (Illustration 7.1). As we established in Chapters 2 and 3, all contem-

ILLUSTRATION 7.1 Archaeological site protection, Murramarang, New South Wales, Australia, aims to stem the tide of change, in this case from coastal erosion

porary landscapes are the product of long-term environmental processes in which change is the norm. Global change analyses are predicated on the continuation, indeed acceleration, of change in the future. What sort of management is appropriate in this context?

We shall discuss here both landscapes valued mainly for their physical and biotic features, and those which can be termed built landscapes. This approach tries to get away from the nature/culture dualism that was critiqued in the first half of the book. It also recognizes that the issues of change relate to both, albeit often on different timescales. A built structure, like an old-growth forest, is produced in a particular set of circumstances. Neither is frozen in time; they are subject to social and ecological processes operating at a range of scales. The built structure may be used, abandoned, subject to decay, reoccupied, and partially rebuilt over hundreds or thousands of years. The composition and structure of the forest changes as climate changes, hunter-gatherers use it, and it is subjected to storm damage or disease over similar timescales. Fire of different intensities and frequencies will differentially affect different forest components. In both cases what survives for us is a complex product of all these processes. In both cases there may be good reasons for wanting to protect the landscape from damage, but in neither case can we protect it from change. An important thread throughout this chapter, then, is the necessity to clarify what we can and should protect in different contexts. Informed environmental management requires us to be more explicit about our goals. We have the tools to do this, as the examples in this chapter will show.

Thinking about environmental change is inseparable from the dimension of time. Indeed, the key terms in this chapter contain an implied temporal direction. *Restoration* contains the idea of going backwards, of returning to a previous condition, whether of forests, rivers or buildings. *Preservation* implies maintaining some element of present conditions or processes. *Creation* looks towards the future. Important notions of time and change are embedded within the termi-

nology of landscape management; I will not treat this as a simple definitional question that needs a certain amount of clarification before we get to the real business. Rather, in this chapter it is important to explore these cultural understandings in order to clarify the basis of decision-making. As Higgs (1997: 339) argues, 'failure to achieve clarity on moral and cultural considerations will hinder ecological restoration's potential to generate healthy relationships between the people and the land'.

7.2 Restored landscapes: going backwards?

The changing cultural context is seen clearly in, for example, the definitional wrestling over ecological restoration, as reviewed by Higgs (1997) (Table 7.1). Past conceptualizations of restoration as 'applied succession' (Parker, 1997) show how closely it is related to a particular view of ecosystem change. Higgs argues that good restoration requires attention to the cultural context of both the restoration process itself and the environment being restored. Restoration is something 'constructed by human values and attitudes' (1997: 345). The cultural history will be crucial to answering questions such as 'To what conditions should this environment be restored?' (e.g. Illustration 7.2). Challenges to the concepts of 'indigenous' or 'original' ecosystems have come from both the Old World, with long histories of intensive usage, and indigenous people, who also have a long history of use. Ecological fidelity is thus a necessary, but not a sufficient, component of restoration. Human dimensions are important in the three principles of ecological fidelity:

- *Structural/compositional replication of the so-called natural ecosystem.* 'For example, a tallgrass prairie must have close to the expected diversity of floral and faunal species' (Higgs, 1997: 343). This is likely to require not just controlled burning but much more detailed activities including weeding, selective plantings, use of herbicides, thinning and, importantly, time.

TABLE 7.1 Definitions of ecological restoration, and associated issues/problems

Organization, date	Definition	Issues/problems
Society for Ecological Restoration (USA), 1990	The process of intentionally altering a site to establish a defined, indigenous historic ecosystem. The goal of this process is to emulate the structure, function, diversity and dynamics of the specified ecosystem.	Why is one time-slice preferable to another? How practical where evidence of past ecology is erased? What does indigenous mean in areas with long periods of human interaction?
Society for Ecological Restoration, 1995	The process of renewing and maintaining ecosystem health	Very general Of little use in describing basic elements of restoration practice
US National Research Council, 1992	The return of an ecosystem to a close approximation of its condition prior to disturbance. In restoration, ecological damage to the resource is repaired. Both the structure and the functions of the ecosystem are re-created. ... Often ... requires ... reconstruction of antecedent physical hydrologic and morphological conditions; chemical clean-up or adjustment of the environment ...	Technically detailed Attends to balance between function and structure No cultural context – what (and when) is disturbance?

Source: After Higgs (1997)

- *Functional success.* 'The ecosystem must align ecologically with the system it is designed to reproduce' (p. 343). This usually depends on management, e.g. use of fire on the prairies. 'Also, some ecosystems depend on human practices – for example, cultural landscapes that have long involved humans as part of the ecological functioning' (p. 343).

- *Durability.* The restoration 'must hold up over a significant period of time, significant being defined relative to the type of ecosystem' (p. 344).

Central to thinking about restoration is the question of natural states. Is there a baseline to which systems can be returned? Following their important role in the articulation of non-

ILLUSTRATION 7.2 Minnamurra National Park, New South Wales, protects a small area of rainforest, much of which has regrown following clearance for agriculture in the nineteenth century. Interpretive material in the park glosses over these processes of historical change

equilibrium ecology (*see* Chapter 3), Pickett and Parker see one of the pitfalls of restoration ecology as the belief 'that there is one reference state or system that can inform restoration' (1994: 75; *see also* Hobbs and Norton, 1996). The second pitfall is to see it as a discrete event rather than intervention in an ongoing process.

> Contingency means that restoration ecologists will have a variety of reference states to choose from. Contingency establishes a whole range of systems, not just one 'climax' or predisturbance state. Of course there are many ecological and societal reasons to choose certain reference states, including aesthetics, commodity production, ecosystem services, and species protection, among others. But the point is that restoration ecologists must choose, and nature provides a range of ecologically valid system states. (Pickett and Parker, 1994: 76)

Others reply that some standard or reference is necessary to evaluate the success of a restoration project (Aronson *et al.*, 1995), particularly when many practitioners of restoration are not immersed in the nuances of the ecological literature. The agreement here is that restoration is a matter of making conscious human choices about intervention.

Aronson *et al.* (1995) emphasize the importance of a reference system's being indigenous and historic. While the idea of alternative future states includes a temporal dimension in that direction, the past history is not always considered (Hobbs and Norton, 1996; Aronson

and Le Floc'h, 1996) (Figure 7.1). Processes in past time are felt to be particularly important in order 'to ascertain whether any "thresholds of irreversibility" have been crossed' (Aronson and Le Floc'h, 1996: 330), and to discern the amplitude of change (Figure 7.2).

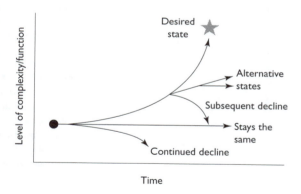

FIGURE 7.1 'Traditional view of restoration options for a degraded system, illustrating the idea that the system can travel along a number of different trajectories and that the goal of restoration is to hasten the trajectory towards some desired state. In this view, the past history of the system is not considered, yet the route by which the system reaches the present point can have a large impact on the potential for restoration'
Source: Hobbs and Norton (1996: figure 2). By permission of Blackwell Science Inc.

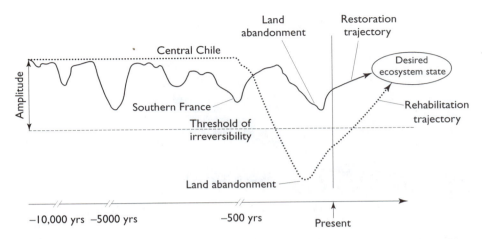

FIGURE 7.2 The Hobbs and Norton figure after modification: the comparison of two specific areas with similar Mediterranean bioclimates. We have added a distinction between restoration and rehabilitation, and we emphasize the notion of trajectories . . . . Declining curves correspond to periods of increasing human exploitation of resources and, often, increasing human density'
Source: Aronson and Le Floc'h (1996: figure 2). By permission of Blackwell Science Inc.

7.2.1 Restoring pre-human New Zealand forests

As a means of clarifying these issues, we begin with an argument that says we can and should go backwards. New Zealand, argues palaeoecologist Matt McGlone, was colonized by humans so recently that it is possible to aspire to a pre-human baseline. While acknowledging the extent and causes of change in Australasian terrestrial ecosystems – megafaunal extinctions, introduction of exotics, soil degradation through agriculture, and increasing levels of atmospheric carbon dioxide – McGlone (1999) argues that this enhances rather than undermines the rationale for restoration. For both biological (genetic potential) and moral (their right to exist) reasons, we want 'self-sustaining, resilient ecosystems capable of responding to future environmental change' (McGlone, 1999: 3). The key to this, he argues, is a 'natural environment template [that] summarizes what past environments were like, what present environments could be like without gross human interference, and how they might be in the future' (p. 3). A template is seen not as a static baseline, but as a multi-layered work in progress that can be modified according to new data and ideas.

For New Zealand the most important level is the pre-human state, about 800 years ago, still within the lifespan of some of the older forest trees. A second level, the AD 1840 horizon (date of the Treaty of Waitangi, symbolizing the conversion from sole Maori management to British control), 'is currently used by the New Zealand Department of Conservation as a *de facto* starting point for considering landscape issues' (McGlone, 1999: 6). McGlone considers use of this time as a baseline to be problematic because it interprets large areas of non-forest ecosystems – scrub, grassland and fernland – as indigenous and of prime conservation significance. For McGlone, however, the 'natural' New Zealand landscape is 85 per cent forest, a goal attainable 'in the absence of constant human effort to arrest ecological successions' (p. 6) (Figure 7.3). As rural land is increasingly abandoned, the natural template will be essential to make sense of changing ecosystems (Box 7.1).

I am engaging with McGlone's arguments at some length here because he is one of the few palaeoecologists to go beyond the rhetoric that the past has relevance for the future. He has attempted to map out explicit goals for the

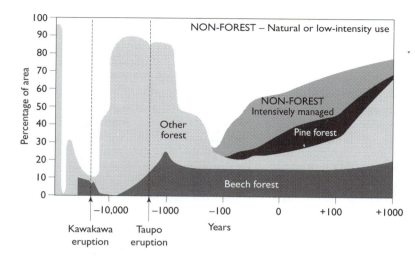

FIGURE 7.3 Past and future vegetation cover of New Zealand
Source: McGlone (1999). By permission of Matt McGlone

future, based on the past. Even for Australia, with a much longer human history than New Zealand (*see* Chapter 2 for an overview of these debates), McGlone thinks 'it will be important for Australians to have the "without-human template" to remind them what natural ecosystems can be without humans' (1999: 6). While I can see this as a useful device for imagining other possibilities, to escape the tyranny of the present (Box 7.1), the climatic conditions of pre-human Australia were very different from those of the present. I am less convinced

---

Box 7.1 Using the natural environment template to make sense of changing ecosystems in New Zealand

Preventing the tyranny of the present

Reconstruction projects are focusing on 'remnant indigenous species left not from an original ecosystem, but from a previous human induced ecosystem', 'simply because the original ecosystems have been reduced to near invisibility' (p. 7). The Maori-induced ecosystems are deceptively natural, although 'often celebrated and cared for as unchanging fragments of lost landscapes' (p. 7).

Establishing a time frame

If we accept the need to plan for the long term, how long should it be? For McGlone, the only really relevant timescale is the glacial–interglacial cycle, about 120,000 years. 'All the extremes for which a species has adapted or can withstand, will occur within this cycle, and this enables prediction of the spatial scale needed' (p. 8).

Establishing a spatial scale

Different ecosystems have different needs. Some very specialized ecosystems and their species may survive, indeed evidently have survived past cycles, in small landscape fragments. 'For others – such as most forest ecosystems – freedom to contract and expand in relation to landscape disturbance and climatic change is essential in the long term' (p. 9).

Security and planning for disturbance

Particularly in the New Zealand context, major disturbances (earthquakes, volcanic activity) are part of the pattern of change. An historical perspective is essential to understand the return time and spatial scale of these.

Source: McGlone (1999)

(less optimistic?) than McGlone that the glacial–interglacial cycle is the relevant timescale to aim for.

The attempt to shift the focus of attention and management from 'the lightly modified protected estate' to the 'vast areas that have been heavily or almost completely modified by human activities' (p. 4) is of relevance to many New World contexts beyond Australasia. Similarly, McGlone's critique of patchy 'theme park' ecosystems within an intensively managed landscape reminds us that preservation and conservation resources are often misdirected. If he is right that neither Australia nor New Zealand will depend on wide-scale agriculture and forestry for much longer (at least within the glacial–interglacial time frame!), then large areas of land will undergo changing use. In contrast to the New Zealand situation, there would be few Australian ecologists who would expect a 'succession' to forest cover to ensue; many argue that the Australian landscape has become so degraded through agriculture that irreversible thresholds, of for example soil fertility, have been crossed. For all that they are lumped together in northern eyes as 'the Antipodes', Australia and New Zealand are very different ecologically and in their human history, illustrating that these issues need to be worked through in their local context.

McGlone's arguments about naturalness raise another set of issues. While an important thread of his argument is that many New Zealanders have treated heavily modified Maori ecosystems as unproblematically natural, he in turn evokes an unproblematically natural pre-human past. Even assuming such a restoration is possible, it is difficult to see how this sits with a human presence that is by definition 'unnatural'. Under such a definition, human activity can only be damaging; there appears no conceptual space here for creative human activity. And yet, as McGlone acknowledges, people are going to want to live in these recovering forest areas. With the marketing of New Zealand as a 'natural' tourist destination, the pressure from visitors is also set to increase.

McGlone comes closer to solving this apparent paradox when he explains that the problem with modified ecosystems is not that they are unnatural *per se*, but rather that they are floristically impoverished (1999: 5). While this is an important point, and one that can be discussed on its merits, there are a number of implicit assumptions here about past and present Maori impacts and aspirations. As with other indigenous peoples, Maori perspectives on conservation are rarely encountered in the scientific literature (Roberts *et al.*, 1995). These issues will be discussed further in Chapter 8.

A further interesting issue is the assumption that climate change will be the main driver of environmental change in the future. This may be the case for a New Zealand with a shrinking economy, but it is not likely to be the case in most parts of the world with rapidly growing populations (Table 3.2, p. 44). There is a danger here in assuming that the needs of urban dwellers can be met independent of the increasing areas that fall under the ecological footprint.

7.2.2 Restoring traditional agricultural landscapes

In north-west Europe, with its much longer period of human history, the issue of landscape restoration is conceptualized quite differently. In Britain and France, for example, there is a focus on restoring the traditional agricultural landscape, characterized by various combinations of hedgerows, small fields, stone walls and a patchwork of wooded and meadow areas (Illustrations 7.3a and 7.3b). Degradation of this landscape is relatively recent, for the most part a feature of intensive post-Second World War agriculture, accelerating since the Common Agricultural Policy (CAP) of the European community (now the European Union). Estimates of hedgerow loss are as high as 192,000 km for England and Wales (1946–74) and 6000 km for the Grampian region of Scotland (1940s–1970s) (Ghaffar and Robinson, 1997: 211). 'In effect, the agricultural landscape has been the evolving stage upon which the drama of the CAP's price support policies has unfolded via a technologically sophisticated drive for increased productivity' (Ghaffar and Robinson, 1997: 205).

(a)

ILLUSTRATION 7.3 (a)
Agricultural landscape with hedgerows, West Lothian, Scotland (by permission of Gordon Waitt)

(b)

ILLUSTRATION 7.3 (b)
Land ploughed to edge of hedgerow, East Lothian, Scotland (by permission of Gordon Waitt)

In their study in south-east Scotland Ghaffar and Robinson utilized aerial photographs, questionnaires, historical maps and a geographic information system (GIS). Landscape change in four parishes was seen in both field size and field boundaries. Average field size increased 21.4 per cent between 1972 and 1988. Changes in field boundaries were more variable over this same time period, reflecting attempts at restoration which commenced in the 1980s (Table 7.2 and Figure 7.4). Interest in hedgerow restoration arose out of concern for habitat of birds and small mammals, as well as the attractiveness of the landscape. Hedgerow replanting has been at the expense of easily removed post-and-wire fencing and was cheaper and less labour-intensive than stone wall reconstruction (Table 7.2).

There are shifts occurring which need to be considered at several scales. East Lothian reflects broader trends of hedgerow restoration; 25,600 km of hedges were replanted in England between 1984 and 1990. Overall, however, more hedgerows (85,000 km) were removed in England in that period than replanted (Ghaffar and Robinson, 1997: 214). In this respect the CAP has been an agent of

TABLE 7.2 Percentage changes in field boundaries, 1972–88, in four parishes in East Lothian, Scotland

Parish	Hedgerows	Post- and-wire fences	Stone walls	Vegetative belts	Tree line	Ditches and others
Athelstaneford	+5.3	−10.2	−6.5	+4.6	−1.5	+0.1
Haddington	+9.3	−18.7	−4.3	+4.2	−11.0	−4.4
Morham	+41.3	−26.8	0	+3.4	+4.6	−17.5
Prestonkirk	+20.0	−15.4	−11.7	+8.0	+2.4	−11.1

Source: Based on aerial photographs summarized from Ghaffar and Robinson (1997: Table 6)

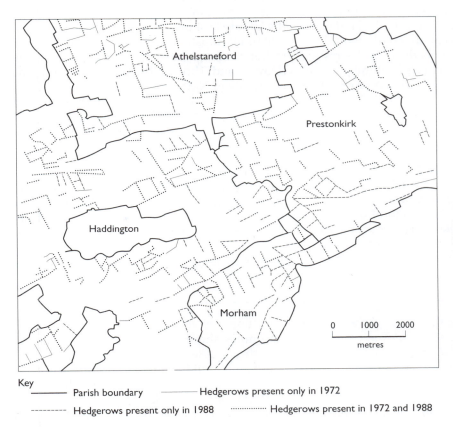

Key

——— Parish boundary ——— Hedgerows present only in 1972

--------- Hedgerows present only in 1988 ············ Hedgerows present in 1972 and 1988

FIGURE 7.4 Hedgerow replanting, East Lothian, Scotland, 1972–88, based on aerial photographs *Source*: Reprinted from *Geoforum* **28**, A Ghaffar and G. M. Robinson, Restoring the agricultural landscape: the impact of governmen policies in East Lothain, Scotland, pp. 205–17. Copyright 1997, with permission from Elsevier Science

change in both directions, with the European Union now contributing funds to a number of restoration schemes. In Lower Normandy, France, residents showed a willingness to pay a reasonable amount each year for the restoration and replanting of hedgerows (Bonnieux and Le Goffe, 1997). Ghaffar and Robinson argue that various schemes of environmental restoration, including the designation of Environmentally Sensitive Areas (ESAs), make farmers 'producers of countryside and landscape and not just suppliers of food'. However, 'farmers' responses to the schemes are extremely varied, resulting in widely differing patterns of uptake and impacts on the landscape' (p. 215).

ESAs, now designated in Ireland, Denmark and the Netherlands as well as the UK, are increasingly likely to influence the shape of future European landscapes. In the Scottish programme, farmers gain an annual payment 'in return for following a prescribed set of farming practices designed to protect and enhance the environment' (Simpson *et al.*, 1997: 307). Simpson *et al.* made visual predictions in two contrasting Scottish landscapes – Breadalbane in the southern Highlands and the Machair of the Western Isles – with and without conservation policy as contained in ESA prescriptions. They found that the policies will lead to landscapes 'that maintain and increase diversity of land cover and linear features' (p. 317). Without the policies, both landscape structure and botanical diversity would be simplified. Although embedded in desires for 'traditional' landscape and agricultural practice, the central contradiction of the ESA programme is that it depends on non-traditional activities such as fencing off areas for revegetation.

7.2.3 River restoration

River restoration is also most developed in north-west Europe, particularly the UK, and faces many of the same issues as restoration ecology, as Brookes's discussion of definitions shows (Box 7.2). Brookes sees the problems of determining 'pre-disturbance' conditions as practical rather than theoretical ones, and the

relevant timescale being decades and centuries rather than longer. Yet it is clear from long-term studies that rivers change at a range of spatial and temporal scales (Brown and Quine, 1999); the concept of stability being a natural condition has become just as problematic in fluvial geomorphology as in ecology. Further, because floodplains are 'attractors' of human settlement (Brown, 1996), humans have been part of the riverine environment for thousands of years (Brown, 1997b).

In practice, most river restoration projects in the UK are best understood as involving 'rehabilitation' (Illustration 7.4) or habitat 'creation' rather than a return to original conditions (Downs and Thorne, 1998). For example, the River Idle in Nottinghamshire was extensively channelized in the early 1980s for drainage and flood mitigation, contributing to low ecological diversity, a poor fishery and low aesthetic value. Rehabilitation provisions included planting in the river corridor; deflectors and riffle installation to increase morphological diversity in the stream channel; and reed planting and fish shelters to enhance habitat (Downs and Thorne, 1998). In addition, catchment-wide measures to reduce sedimentation were proposed, although these are necessarily longer-term objectives.

Box 7.2 Some definitions of river restoration

- (Cairns, 1991) The complete structural and functional return to a pre-disturbance state.
- (Gore, 1985) In essence, river restoration is the process of recovery enhancement. Recovery enhancement enables the river or stream ecosystem to stabilize (reach some sort of trophic balance) at a much faster rate than through the natural physical and biological processes of habitat development and colonization. Recovery enhancement should establish a return to an ecosystem which closely resembles unstressed surrounding areas.

- (Osborne *et al.*, 1993) Restoration programmes should aim to create a system with a stable channel, or a channel in dynamic equilibrium that supports a self-sustaining and functionally diverse community assemblage.
- (Herricks and Osborne, 1985) Implicit in the concept of water quality restoration is some knowledge of the undisturbed or natural state of the stream system. Restoration of water quality can be defined as returning the concentration of substances to values typical of undisturbed conditions.

Source: Brookes (1995)

ILLUSTRATION 7.4 River rehabilitation: pin groynes designed to accumulate sediment on the bar beneath the opposite bank and prevent erosion of the bank, Taylors Arm, Nambucca catchment, Australia (by permission of Gerald Nanson)

In Germany the 'main objective of stream rehabilitation is the conservation of nature' (Kern, 1992: 324), for which the concept of *Leitbild* is used. *Leitbild* 'is a description of the desirable stream properties regarding only the natural potential, not considering the economic or political aspects that influence the realization of a scheme' (p. 325). It is based on three elements, the second and third of which include a sense of past history:

- natural stream properties (stream pattern, morphodynamics, floodplain morphology, natural flow dynamics and flooding, potential vegetation, etc.);
- irreversible changes of abiotic and biotic factors (e.g. changes of the run-off regime or sediment transport, dredging of alluvial sediments in the floodplain, extinct species, etc.);
- aspects of cultural ecology; specific traditional stages of land use caused an increase of species which are endangered by modern agriculture. (Kern, 1992: 325)

In many of these examples, however, the history at issue is hundreds rather than thousands of years old. Channelization of rivers associated with industrial growth is seen to be the most significant change. In the USA restoration is implicitly towards a pre-European baseline (e.g. Kondolf 1995; Goodwin *et al.*, 1997). Where the relevance of a prehistoric timescale is acknowledged, the

methodological issue of 'how do we know?' is more likely to be engaged with. For example, discussing the Kissimmee River in Florida, Dahm *et al.* (1995: 226) found that 'insights into presettlement [i.e. pre-European] historical conditions, such as the widespread presence of habitat-forming large woody debris (LWD) or snags are also found in written reports of pre-channelized river structure'. For the post-European period, information from archival resources was rendered into GIS maps of river and floodplain change.

7.2.4 The restoration of built structures

Issues in the restoration of built structures offer an important comparison to those of continuity and change within biotic landscapes. Various definitions and guidelines used by the International Council of Monuments and Sites (ICOMOS) are summarized in Box 7.3. Similarities to ecological restoration include the ideal of fidelity to the original integrity of the place or structure, and the recognition of the broader context or setting of the significant structure. Because the significance is explicitly related to human activity, ICOMOS charters are more likely to acknowledge ongoing human interaction with the site. They are also more likely to refer to multiple periods of use or occupation (e.g. Venice Charter, article 11), rather than allude to a single 'original' state or event. The very limited circumstances in which

Box 7.3 Restoration and related concepts as used in International Council of Monuments and Sites (ICOMOS) charters

Venice Charter for the Conservation and Restoration of Monuments and Sites (1964)

- Article 9. The aim [of restoration] is to preserve and reveal the aesthetic and historic value of the monument and is based on respect for original material and authentic documents. It must stop at the point where conjecture begins, and in this case moreover any extra work which is indispensable must be distinct from the architectural composition and must bear a contemporary stamp.
- Article 11. The valid contributions of all periods to the building of a monument must be respected, since unity of style is not the aim of a restoration. When a building includes the superimposed work of different periods, the revealing of the underlying state can only be justified in exceptional circumstances …
- Article 12. Replacements of missing parts must integrate harmoniously with the whole, but at the same time must be distinguishable from the original so that restoration does not falsify the artistic or historic evidence.
- Article 15. Ruins must be maintained and measures necessary for the permanent conservation and protection of architectural features and of objects discovered must be taken …

  All reconstruction work should however be ruled out 'a priori'. Only anastylosis, that is to say, the reassembling of existing but dismembered parts can be permitted. The material used for integration should always be recognizable and its use should be the least that will ensure the conservation of a monument and the reinstatement of its form.

The Australia ICOMOS Charter for the Conservation of Places of Cultural Significance (the Burra Charter) (1981)

- Article 12. Preservation is limited to the protection, maintenance and where necessary, the stabilisation of the existing fabric [all the physical material of the place] but without the distortion of its cultural significance.
- Article 13. Restoration is appropriate only if there is sufficient evidence of an earlier state of the fabric and only if returning the fabric to that state recovers the cultural significance of the place.
- Article 14. Restoration should reveal anew culturally significant aspects of the place. It is based on respect for all the physical, documentary and other evidence and stops at the point where conjecture begins.
- Article 15. Restoration is limited to the reassembling of displaced components or removal of accretions in accordance with Article 16.
- Article 16. The contributions of all periods to the place must be respected. If a place includes the fabric of different periods, revealing the fabric of one period at the expense of another can only be justified when what is removed is of slight cultural significance and the fabric which is to be revealed is of much greater cultural significance.

Appleton Charter for the Protection and Enhancement of the Built Environment (1983)

- Preservation – retention of the existing form, material and integrity of a site.
- Period restoration – recovery of an earlier form, material and integrity of a site.
- Rehabilitation – modification of a resource to contemporary functional standards which may involve adaptation for new use.
- Period reconstruction – re-creation of vanished or irreversibly deteriorated resources.

- Redevelopment – insertion of contemporary structures or additions sympathetic to the setting.

Source:
http://life.csu.edu.au/~dspennem/VIRTPAST/Conventions/ICOMOS

reconstruction is permitted illustrate that this is not an attempt to return to the past. Perhaps the most important difference is that in cultural heritage, the restoration process must explicitly draw attention to itself, and label itself as such (e.g. Venice Charter article 9, Burra Charter article 19). Article 16 of the Venice Charter requires the restoration process to be precisely documented, and the records archived. In other words, by drawing attention to itself as a cultural process, this restoration does not pretend to be 'original' or 'natural'.

These issues are illustrated in the story of the rebuilding of the Globe playhouse in London (Mulryne and Shewring, 1997b). The original Globe (1599–1613) is inextricably associated with the plays of William Shakespeare, and the rebuilding project is of a cultural landscape in the fullest sense, with many layers of meaning embedded in the timber framing and the plays upon its stage. If a visit for the modern audience is 'an exercise in double vision, present and past' (Mulryne and Shewring, 1997a: 15), the original building itself 'sent forth a double message about the past and the present' (Keenan and Davidson, 1997: 147). While the twentieth-century reads the circular playhouse shape as Elizabethan, in the 1590s such structures evoked classical Roman antiquity, in which tradition the theatrical performance was placed. In the rebuilding process, then, the structure was 'dedicated to theatrical experimentation and revaluation, securely grounded in the later twentieth-century' (Keenan and Davidson, 1997: 148). There was painstaking attention to the physical aspects of reconstruction, such as timber-framing techniques, thatching and internal decoration. On the other hand, the design and construction had to contend with gaps and silences in the historical and archaeological evidence, and the demands of today's safety and building standards. For the proponents of the project the iterative process of craft-based construction mirrored the experimental nature of the theatre that had been and would be performed there, and continued a tradition of interplay between past and present. Some of the public, however, had other ideas:

> While nobody involved with the 1990s Globe would claim that it is any more than a shadow, a representation, a laboratory for the player and the theatrical historian, there can be no doubt that many visitors will impose their own notion of reality, their own idea of 'secular pilgrimage' upon it. In the summer of 1996, it has been reported that spectators in Elizabethan costume were turned away from a (modern-dress) production of *The Two Gentlemen of Verona*. Before the structure was even finished, it had been appropriated, theme-parked (by its visitors) and (at the same time) reified, all in direct contradiction to the stated aims of those who have worked for the project and on this book. (Keenan and Davidson, 1997: 147)

If people do this to an explicitly and reflexively cultural production, how are they going to interpret ecological restoration which does not draw attention to itself? Will a patch of planted trees be interpreted as a natural forest, and does it matter? For example, in New Zealand, the majority of sphagnum bogs valued as pristine remnants are less than 150 years old, the product of hydrological changes caused by intensive post-European burning and grazing (Wilmshurst and McGlone, 1998). While this does not mean they should not be valued, appropriate management strategies cannot be developed from the wrong history.

7.3 Preserved landscapes: freezing the present?

Both philosophical and practical dilemmas of preservation (Illustration 7.5) are exemplified by the situation of rock art. Painting and engraving of rock surfaces is an iterative process, building up over periods as long as

Illustration 7.5 'Williamsburg, Virginia: the main street with inn and horse-drawn tourist wagon in a "frozen" streetscape.' (by permission of Graham *et al.* (2000: 102))

tens of thousands of years. Over those timescales weathering, erosion, abandonment and reworking are all involved. The rock art story is thus inherently one of both change and continuity. Increasing interest in rock art from heritage and tourist perspectives is couched in terms of preservation and protection, from both biophysical processes and visitor pressures. The often brutal process of colonization brought an abrupt end to living traditions of rock art in many areas, exacerbating perceptions of it as being frozen in time. Although it is recognized that 'the single largest threat to heritage sites today is the people who visit them' (Jacobs and Gale, 1994: 131), rock art will naturally decay in the absence of people. (Some of the physical processes involved are summarized in Box 7.4.) Conservation measures are thus necessarily interventionist, even if they can at most slow down the process of decay. The most common and least intrusive measure is the use of silicone driplines to divert surface water run-off from the art. The use of physical and psychological measures to reduce visitor pressures will be discussed further in Chapter 9.

Circulating through any discussion of

Box 7.4   Rock weathering processes affecting art in Clarens Formation sandstones, South Africa

- Rock weathering processes are affected by the mineralogy, rock structure, rock properties and particularly the rock moisture and thermal regimes.
- The principal rock weathering mechanisms are thermal stress fatigue; salt crystallization; hydration and dehydration of rock minerals, clay minerals and precipitated salts; solution; hydrolysis and chemical alteration.
- The rock moisture regime is the most important environmental control on the weathering processes.

- Rock weathering processes cause granular disintegration and the enlargement of pores and bedding planes.
- Although there is more than one temporal weathering regime, those most damaging to the rock art are the short-term changes that affect particularly the rock surface and the area immediately beneath it.
- Rock weathering processes are accelerated by existing weathering that produces a more dynamic environment with respect to both the thermal and rock moisture regimes.

Source: Wahl (1999: 5), from Meiklejohn (1995)

restoring and preserving landscapes at the broader scale are different traditions of public and private space and ownership. In New World situations such as the USA, Australia and New Zealand, preservation is usually equated with reservation – the setting aside of public land in forms such as national parks and reserves. In these colonial contexts, public land has been thought of as empty land. Indeed, it is often joked among conservationists that the land available for national parks is the land that is no good for anything else, and of no interest to anyone else. This view of potential reserve land as a blank slate, as *terra nullius*, is now being challenged by indigenous people, who point out that they have never gone away, although their presence has been rendered invisible by various cultural processes. (These issues will be discussed further in Chapter 8.) This notion of reserved lands being empty of people is quite different from Old World situations, where a long period of human interaction has been impossible to ignore. In the UK, for example, although national parks and nature reserves are important, they are part of a mosaic of public and private land to which there are long traditions of access (Thomas *et al.*, 1997). Although this is not the place to discuss in detail land tenure or the different legislative mechanisms of protection, it is important to be aware that these issues take on increased relevance in particular local situations.

Establishment of reserved lands is an important mechanism of habitat and biodiversity protection, particularly in areas of high population growth and increasing land-use pressure. These include situations as different as the relatively small island of Mauritius (Safford, 1997) and China (Jiantian, 1998). However, reservation is not a panacea, and it must be considered within both spatial and social context. Designation of marine parks and reserves in Hong Kong will not necessarily protect local biodiversity from the impacts of intensive urbanization (Jian-Hua and Hills, 1997). Paraguay, for example, has moved within the past 30 years to identify potential reserved land, and the proportion of land in national parks and reserves (3.59 per cent) exceeds both the South American average (2.7

per cent) and that of the USA (2.2 per cent) (Yahnke *et al.*, 1998). However, a lack of research means that the effectiveness of these areas in preserving mammal biodiversity is not yet well understood. Yahnke *et al.* argue that while landlocked Paraguay does contain rare habitats such as remnant Atlantic coastal forest and dry Chaco (a mosaic of grasslands, palm savannas, open woodlands and xeric thorn forests), research on these has suffered because of the focus on the tropical rainforests further north.

The so-called Yellowstone model of national parks, which contain relatively 'pure' public land and are contrasted with the degraded private land of urban spaces, is very much a cultural construction, as we discussed in Chapter 4. It is also quite specific to frontier societies such as the USA and Australia. In most parts of the world the boundaries of reserved lands are more physically and socially permeable. Reserved lands often include residential human activity and subsistence land use of some sort, as opposed to a more recreational focus. Conversely, this permeability illustrates the importance of what is going on outside the boundary.

It is increasingly recognized that reserved lands are for the most part too small, too fragmented and too unevenly distributed to provide a framework in which biological processes can continue as normal, let alone respond to accelerated environmental change. The need for corridors to link habitat islands, or stepping stones between refugia, means that combinations of public and private land, across different types of land use, are required. In Tanzania, mammal extinction, even in very large park complexes such as Serengeti National Park–Ngorongoro Conservation Area, is likely to continue unless corridors can be established (Figure 7.5). It has been estimated that ongoing expansion of cultivated areas limits the window during which this could be done to about five years (Newmark, 1996).

Issues of fragmentation, patchiness and aggregation need to be considered in both ecological and sociopolitical terms. Park effectiveness is a matter not just of size, but also of shape and spacing. In an assessment of

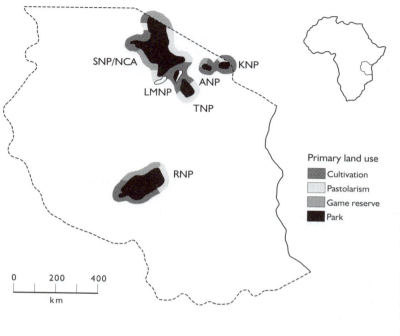

FIGURE 7.5

Location in Tanzania of Kilimanjaro (KNP), Arusha (ANP), Lake Manyara (LMNP), Tarangire (TNP), Ruaha (RNP) national parks, and Serengeti National Park/Ngorongoro Conservation Area (SNP/NCA), and primary land use immediately adjacent to these parks
Source: Newmark (1996: figure 1)

national parks throughout Africa, Siegfried *et al.* (1998) argued that southern and south-central African parks are most at threat from edge effects, owing to their large edge to interior ratios. Park separation is greatest in northern Africa, with an average inter-park distance of 2396 km. Both these measures have implications for biotic migration possibilities, and for human activity immediately adjacent to the park (*see* Chapter 8). The recognition that environmental issues straddle political and cultural boundaries has led to proposals for 'superparks' involving the cooperation of several adjacent nation-states. One proposed between Zimbabwe, South Africa and Mozambique, encompassing existing national parks including Kruger, would need to be funded largely by tourism to pay its own way (Duffy, 1997) (Figure 7.6). The proposal has met resistance for a complex of political reasons, including worries about the exclusion of local people, fears of loss of sovereignty, the dominance and influence of international aid donors and land distribution issues (Duffy, 1997). The ecological goal of reopening wildlife migration

routes cannot be met without addressing these issues.

Given the great variability in the extent and status of reserved lands, and the range of pressures on non-reserved lands, what are the implications for landscapes and their biotic components to future environmental change? There is an extensive body of literature attempting tests under different scenarios of future climate change, themselves the subject of debate and controversy. It is also important to remember that consideration of climate change should not deflect attention from the ongoing agency of human activity. For example, a study in Mexico suggests that for ecosystems such as tropical forest, projected climate change is much less of a threat than clearance for cattle ranching (Villers-Ruiz and Trejo-Vazquez, 1998).

Partly because of the complexity of contributing factors, most simulations of plant response to climate change are projected onto a natural landscape which exists only in the imagination of the computer. Consider the example of *Abies alba* (silver fir), *Acer pseudopla-*

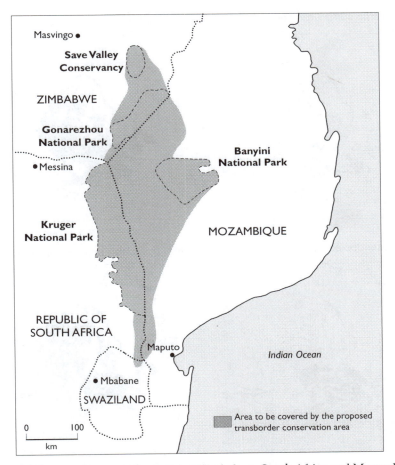

FIGURE 7.6 Proposed superpark spanning Zimbabwe, South Africa and Mozambique
Source: Duffy (1997: figure 1)

tanus (sycamore) and *Quercus ilex* (evergreen oak). Their natural distribution in north-west Europe is modelled under the present climate, and all are candidates for northward movement under a global warming scenario (Figure 7.7) (Sykes, 1997). Neither model maps where these tree species really are now, or where they would be under a changed climate, because of human impacts on both those distributions. This is of course recognized by the author in the caption – 'potential' distributions – and in the accompanying text. Crucial to the effective interpretation and use of modelled scenarios is the recognition that none of these species could achieve those distributions independent of human activity. As Sykes argues,

northward migration is hampered by the heavily modified landscape, but could be facilitated by plantations. *Acer pseudoplatanus* could spread rapidly out of garden cultivation.

These and other simulations confirm the palaeoecological findings, discussed in Chapter 3, which show that species migrate individualistically rather than as a community in response to climate change. What do the two bodies of research together suggest about the relevance of the past record to future landscape change? While the question ultimately needs to be answered at the regional and local scales, some broad principles are summarized in Box 7.5.

Abies alba current climate

Abies alba mean GCM climate

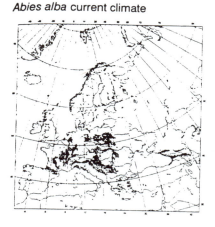
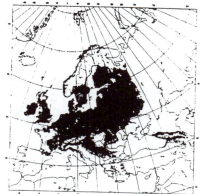

Acer pseudoplatanus current climate

Acer pseudoplatanus mean GCM climate

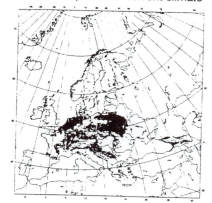
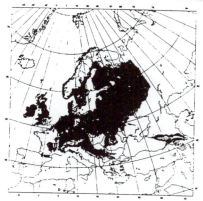

Quercus ilex current climate

Quercus ilex mean GCM climate

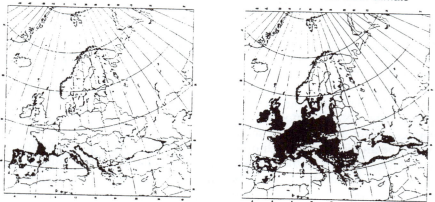

FIGURE 7.7 Simulated potential distributions of *Abies alba*, *Acer pseudoplatanus* and *Quercus ilex* under the present climate and a mean $2 \times CO_2$ climate as predicted by four atmospheric general circulation models (GCMs)

Source: Sykes (1997: figure 2). Copyright Springer-Verlag 1997

Box 7.5 Biotic response to future global change: predictions from the past

- Organisms will not be able to achieve evolutionary adaptation on the timescale of forecast changes (1–2 centuries)
- Some fragmentation, population reduction, loss of genetic diversity and extinction will occur.
- Biomes will not shift as intact entities because species respond differently. Thus new combinations of species will arise.
- Large-scale spatial responses may dominate in some regions (e.g. Europe and eastern North America), but elsewhere the scale will be smaller and/or may be less important than shifts within diverse landscapes (e.g. New Zealand, low-latitude mountains)
- Large-scale spatial responses are unlikely to be rapid enough to keep up with the changing environment, in the absence of human intervention. Some plant species can migrate fast enough to keep up with projected climatic change, but only if they can migrate through continuous, relatively undisturbed ecosystems. Some species whose dispersal ability is artificially enhanced by humans may be able to migrate quickly enough to remain in equilibrium with changing climate.

- Regional prediction is difficult because of spatial variability and individualistic response of taxa, but high-risk areas include high-altitude retreats, coastal areas, isolated oceanic islands, and continental 'cul-de-sacs', e.g. northern tundra.
- Global warming events in the Quaternary can provide lessons for biodiversity conservation.
- Appropriate conservation strategy varies regionally.
- Overall characteristics of taxa vulnerable to extinction can be identified, but not particular taxa themselves.
- Presently rare species may be able to exploit disturbed (whether by humans or climate) habitats.
- Presently abundant species may be restricted by climate change and/or changing human pressures.
- Disturbances (e.g. fire, dieback due to insect attacks, invasion of alien species) are increasing in some regions, leading to increasing mortality and more systems in early successional states.
- Most modelled shifts in vegetation distribution were driven by changes in climate, with limited influence from CO_2, except in the arid regions. However, direct physiological effects (e.g. response to elevated CO_2) can affect species' competitive ability.

Source: Summarized from Huntley *et al.* (1997) and Walker *et al.* (1999)

7.4 Created landscapes: a step forward?

In this final section of the chapter I introduce some examples that help further blur the boundaries between natural and cultural landscapes, between preserved and restored systems, and challenge the linear temporal thinking that much of our vocabulary of change forces us into. An important point here is that it is necessary to be much more explicit about landscape management goals; it is insufficient to allude to the notions of preservation and restoration without explaining them and considering them in context.

Some 500 km² of the English Midlands has recently been designated as National Forest – symbolically and physically in 'the heart of England'. Within this area tree cover is to be increased from 6 per cent to 30 per cent within 30 years (Bell and Evans, 1997), requiring some 30 million trees to be planted in configurations including commercial plantations, farm woodlands, urban forests and roadside planting (Cloke *et al.*, 1996). The rhetoric surrounding this announcement locates the forest as both a national and an international icon. Nationally, it expresses the enterprise culture, and is presented by the Countryside Commission as providing a millennial symbol of hope for the nation. 'In providing the enduring qualities of stability and permanence in a rapidly changing world, the project reinforces the symbolic and material significance of trees and woodlands in British culture' (Bell and Evans, 1997: 259). Internationally, it is seen as contributing to Britain's global environmental responsibilities following the Rio de Janeiro Earth Summit in 1992. The designated area includes towns and villages, ancient woodland and farmland, a patchwork of private land and areas of industrially damaged landscapes. While it was thus suitable for the rehabilitative ambition of 'greening the world', execution of the vision required the support of diverse local populations to be sustained.

The process of gaining that support is documented in some detail by Bell and Evans (1997) and Cloke *et al.* (1996). It will not surprise readers of this book that the creation of this landscape is a complex cultural process:

> As a creative, rather than solely a conservationist, strategy, its ambitious ideals link closely with broader political, scientific, and popular debates on economic investment, social improvement, and environmental enhancement. Presented as a contribution to global environmental sustainability, the planning design builds upon supposedly distinctive national characteristics, notably a renewed respect for cultural and ecological diversity, and a commitment to popular participation and to a spirit of enterprise. As cultural identities are being recast within Britain, the aim is that the Forest initiative should harmonise national and international interests with specific regional concerns. (Bell and Evans, 1997: 274)

Whether we create new landscapes or preserve or restore elements of old ones, we do so through cultural means, expressing cultural meanings. Explicit engagement with these issues is important in both cases, leading to clearer definition of goals and better outcomes.

7.5 Implications

- The vocabulary in which we currently express landscape management carries within it temporal references that are not particularly helpful because they imply that it is desirable and possible to return to previous conditions or states.
- The notions of 'natural', 'original' or 'pristine' landscapes are so problematic in this respect as to be of little practical use. More specific goals (e.g. diversity, complexity, vegetation cover) need to be articulated, with the recognition that these will also have cultural dimensions.
- The value of the long-term record is not that it specifies those previous conditions or states but that it can illustrate the processes and mechanisms of change, and the conditions under which they occur, e.g. extinctions. It may also allow the identification of conditions under which irreversible thresholds will be or have been crossed.
- For the purposes of this discussion, all managed landscapes should be understood as cultural landscapes. Policies based on 'preservation', 'restoration' and related terms express certain cultural values, and the mechanisms by which those policies are effected are cultural processes. Failure to recognize this will exacerbate misunderstandings of environmental change.
- Biotic communities need physical and conceptual space to cope with ongoing environmental change.
- Human presence and activity is part of virtually all biophysical landscapes, whether reserved or not.
- Reflection on the human dimensions of the preservation and restoration process is better developed in the cultural heritage than in the ecological literature, and productive interchange can be gained from cross-reading.

8

Humanized landscapes: a place for people?

8.0 Chapter summary

This chapter explores some of the implications of treating people as an integral part of ecosystems. Two main areas are examined: indigenous environmental knowledge and management; and the mostly agricultural issues relating to developing countries. Romantic and dualistic depictions of traditional land use are critiqued. While the prime attachments to land are local ones, larger-scale issues play an increasing role in processes of environmental change in these areas. Examples used include relationships between indigenous and scientific knowledge systems, traditional harvest, fire and legislation, agricultural land adjacent to protected areas, and sacred groves.

8.1 Living landscapes

If we accept that virtually all landscapes have human dimensions, then how do we talk about the role and activities of people in them? This chapter focuses on landscapes in which people live and make a living. We look first at issues of indigenous land use within developed nation states. The main examples come from Australia and New Zealand, but there are parallel debates in many other parts of the world. In some cases these are areas that Western thinking has valued because the human presence was not seen, or was presumed to be of minimal effect – including protected or so-called wilderness areas. In most of these the indigenous presence is a hunter-gatherer one. In the second part of the chapter we look more

at developing countries, and where subsistence agriculture in various forms is the main form of land use.

The issues include the following:

- What are the implications of acknowledging long-term human presence in these areas, in terms for example of indigenous knowledge, indigenous rights and contemporary land use?
- What is the interplay of local, regional, national and global processes?
- In these contexts 'long-term environmental change' includes a variety of social and ecological processes. Many of the environments under discussion have inherently high levels of variability, against which the discernment of mooted future climate change is difficult. They have also undergone substantial environmental change in recent centuries as a result of, for example, colonialism, population growth and population movements.
- Relationships between social and environmental sustainability.
- Multiple use and adjacent use.

We acknowledge much of the Old World as so full – of people, settlement and impact – that it goes without saying that preservation of old landscapes necessarily includes humans. Thus in England, debates are about elite versus mass access to protected areas, or recreation versus hunting, rather than about the presence or absence of people. In much of the New World, erasure of aboriginal presences has been so conceptually strong that we still tend to think of such areas as empty landscapes, thus aspiring to management of wilderness. In fact the commonalities between Old and New are much greater than we have usually recognized. As in the previous two chapters, there are cross-cutting issues of land tenure and designation here. The hegemony of the 'Yellowstone model' of national parks, under which the ideal park is a 'wilderness' preserve not inhabited on a permanent basis, has influenced many colonial countries (Foster, 1992), but is being challenged by indigenous people who object to being written out of history and the landscape. Outside the boundaries of 'protected' areas various combinations of multiple and overlapping uses can be seen. Biodiversity protection cannot depend on protected areas alone; these are too small, patchy and unevenly distributed. In many areas, such as Australia, some endangered ecosystems are mainly or wholly on indigenous land.

8.2 Indigenous peoples and environmental management

Indigenous people are often caught between two stereotypes in these debates: either they are romanticized as the original conservationists, or demonized as hunters of endangered species. The dualism of these views parallels other dualisms we have critiqued in this book: for example, between unoccupied and occupied lands, or between pristine and degraded environments. All have their roots in reconstructed pasts, both human and environmental. To go, then, in this discussion straight to issues such as sustainable hunting without considering the total context in which indigenous landscape management occurs would be premature, and would simply reaffirm the dualism and these essentialisms. We need first to consider how outside views of indigenous land use have been constructed, and how the inside view might differ.

The Batek of western Malaysia call themselves *batek hep*, 'people of the forest' (Lye, 1998). Lye (1998: 2) illustrates the importance of the forest place as 'an enabling mechanism for the continuous production of knowledge', focusing on forest camps (*haya'*) and the connecting network of trails (*harbew, halbew*). Both these important landscape features combine old places, which incorporate memory of past events and peoples, with newly established ones. Movement along the trails facilitates constant engagement with the forest. This patchwork of trails and camps leaves ecological traces of the Batek's history, and in turn shapes people's future experiences of the forest.

Within this context, change is relevant at two scales. At the scale of constant participation in the forest, change is rapid and minutely observed, as disused trails become overgrown, new gaps are opened by animals, and so on. The Batek value both their mobility and their need to live in the forest; both of these are becoming much more difficult with the development pressures on forests, and government attempts to resettle forest people elsewhere.

The Penan of Sarawak are also Malaysian forest dwellers on whose lives logging has had a dramatic effect, but they are much better known to a Western audience because of the involvement of Western environment groups in their campaigns. For many organizations the Penan became icons of resistance to forest destruction (Brosius, 1997). On the basis of his own ethnographic research, Brosius discusses several ways in which complex patterns of Penan life were represented simplistically by the conservation movement. The one of most relevance to our discussion is the way in which concepts of landscape were treated. Brosius's documentation of the role of habitual activity in the construction of Penan landscape knowledge also has interesting parallels with the Batek, although with an emphasis on rivers instead of trails:

> The importance of rivers to the Penan can scarcely be underestimated. In an environment where visibility seldom exceeds 200 ft, these rivers and streams form the skeleton around which environmental knowledge is organized. … When traveling in the forest, Penan are always cognizant of their precise location relative to various rivers. … To Penan however, the landscape is more than simply a vast, complex network of rivers. Above all it is a reservoir of detailed ecological knowledge and a repository for the memory of past events. (Brosius, 1986: 174–5, cited in Brosius, 1997: 59)

In the environmentalist literature the knowledge, and the forest, is transformed. 'For the Penan this forest is alive, pulsing, responsive in a thousand ways to their physical needs and their spiritual readiness' (Davis, 1990: 98, cited in Brosius, 1997: 60). 'To walk in God's forest is to tread through an earthly paradise where there is no separation between the sacred and the profane, the material and the immaterial, the natural and the supernatural' (Davis, 1990, cited in Brosius, 1997: 60). For Brosius, these and related transformations, for example of Penan concepts of resource stewardship, have more to do with the Western romantic tradition than with Penan ideas. They constitute 'a strategy by which a pattern of recognizing landscape and encoding knowledge about that landscape is transformed into an obscurantist, essentializing discourse which in fact elides the substantive features of that knowledge' (Brosius, 1997: 60).

In a sense there is nothing new about this. Primitivist depictions of indigenous people have a long history that goes back beyond nineteenth-century Victorian anthropology (Kuper, 1988; Kuklick, 1993). The extension of similar discourses into conservation debates has been recognized by a variety of authors. For example, Ramos (1994) traces the historical links in the Brazilian Amazon between first European colonization and the 1989 European tour by rock singer Sting and the Kaiapo chief Raoni:

> The reasoning seems to follow a straightforward exercise in western logic: unspoiled nature is pure; the Indian is part of nature; therefore, the Indian is pure. Such purity then becomes associated with the wisdom that the whiteman once had but has lost on his way to technological progress and with it to the destruction of his environment. Now the whiteman badly needs to recover his lost wisdom in order to preserve, no longer simply a nation, but the planet. (Ramos, 1994: 79)

The historical roots of the discourse, and the implicit conflation of contemporary indigenous people with human origins, show that there are complex issues here. There are also important debates about means and ends. We need to tease them out in order to discuss the nitty-gritty of contemporary land management.

In the Penan instance, Brosius argues, both the forests and the Penan are 'endangered' because neither are valued by the Malaysian government. Environmentalists have thus had to show value and make them narratable to both the government and to Western audiences. Penan knowledge thus becomes reduced to 'the sacred or ineffable' and, like the forests themselves, represents 'the last whisper of an ancient past' (Brosius, 1997: 64). This imposes imaginary meanings on Penan knowledge, and obscures others. Further, 'it paradoxically makes generic precisely the diversity that it is trying to advance' (p. 65); thus indigenous peoples become universalized as 'forest peoples', people with 'sacred ecologies', and bearers of universal wisdom (cf. Suzuki and Knudtson, 1992). Obscuring the variable conditions under which indigenous ecological knowledge is defined, produced and transmitted reduces our ability to utilize its insights in anything other than idealistic ways, an issue that is pursued further below.

In contrast to the romanticist position, there is both environmentalist and scientific literature that focuses on indigenous hunting. Again this often pays insufficient attention to the broader context in which usage takes place. There is reference to hunting pressures (Bodmer *et al.*, 1997), endangered species, extinctions past and present, and the desirability of (and difficulty of defining) sustainable use (Bomford and Caughley, 1996). Now all these are important, but, as argued by Langton *et al.* (1999: 3–4), '"conservation", cannot be used in a presumed commonsense way … without bringing within the ambit of the term some of the cultural differences often overlooked in the conservation literature'. Other concepts which must also be examined for their cultural dimensions include traditional or indigenous knowledge systems, resources, traditional rights, and biodiversity itself.

While challenging, and potentially controversial, these issues cannot be ignored. In many parts of the world, it is no longer possible to discuss conservation and land management issues without considering indigenous involvement (Box 8.1). In Australia, for example, there are several reasons – both intellectual and practical – for doing so:

Box 8.1 Aboriginal aspirations and strategies in relation to environmental issues in Australia

Aspirations in relation to environmental issues:

- appropriate Indigenous ownership and management of lands and waters;

- a desire for appropriate legislative recognition of Indigenous resource rights and protection of cultural and intellectual property;

- adequate resourcing of Indigenous conservation initiatives;

- binding agreements in relation to development projects which entrench Indigenous land ownership, long-term economic and social benefits for local communities and achievable standards of cultural and environmental protection;

- greater intergovernmental coordination and better administration of heritage protection regimes to achieve recognition of Indigenous cultural values;

- local and regional management planning for sustainable use, including multiple uses, of terrestrial, marine and coastal resources for the benefit of Indigenous inhabitants;

- some strategies which have been used in Indigenous conservation management and planning:

(a) cooperation of Indigenous knowledge and tradition with non-Indigenous expertise to achieve high-quality assessment, planning, management and monitoring, particularly in commercial utilization of plant and animal products;

(b) development of regional plans, particularly at the catchment level in coastal north Australia;

(c) such regional plans, based on multiple uses of landscapes, are essential for the long-term control of feral animals and plants;

(d) in achieving conservation objectives, traditional practices alone are no match for the rapid population and development of Indigenous territories by the settler state. Indigenous people and their local and regional bodies require collaborative relationships with individuals and organizations in order to meet particular, identified challenges. Success in such collaboration depends on highly qualified and experienced collaborators with a high level of commitment to the integrity of Indigenous laws.

(e) Such measures as these have the potential of overcoming the conflict over resource use which has typified relations between Indigenous, conservation and industry groups.

Source: Langton *et al.* (1999: 114–15)

- 'Land and water subject to Indigenous ownership and governance constitutes a ... substantial proportion of the Australian continent.' Significant proportions of particular ecosystems are under indigenous control, for example arid-zone ecosystems and rangelands.

- 'A proportion of those lands and waters within the Indigenous domain remain subject to Indigenous management systems which have persisted since the late Pleistocene, and include, for instance, the wet tropics and the wet-dry tropics, parts of which are listed as World Heritage Areas and other IUCN categories.'

- 'Within the Indigenous domain, there are Indigenous systems of governance, both customary and contemporary, with significance for the conservation challenges of this area.' These include particularly the knowledge systems built up over many thousands of years. (Langton *et al.*, 1999: 6)

8.2.1 Knowledge and classification systems

Explicit consideration of the relationship between indigenous and scientific knowledge systems has been a theme of research into the fauna of Uluru–Kata Tjuṯa National Park (Baker and Muṯitjulu community, 1992; Baker et al., 1993). The main distinction between the two knowledge systems is that between evolutionary theory and Linnaean taxonomy and the religious and social law of the *Tjukurpa* (*see also* Chapter 6) (Baker and Muṯitjulu community, 1992). For the most part there is great complementarity between the two systems, to such an extent that traditional ecological knowledge can be used as a rapid survey technique (Baker and Muṯitjulu Community, 1992: 185). In general, Aṉangu can frame the scientific knowledge into more of a regional and landscape context. 'For example, Aṉangu noted the importance of claypans associated with acacia shrublands, which provide water for many animals and thereby cause some of them to stay in such areas longer' (p. 185). There are also differences, some expressions of which are summarized in Table 8.1.

Time is a key element of Aṉangu knowledge. In particular it depends on lifetimes of observation, through many different environmental conditions. Maintenance of knowledge and transmission to the next generation requires continuous interaction with country.

Among the profound cultural differences between colonial and indigenous environmental understandings are the different conceptualizations of native and introduced biota. There are interesting parallels here between Aotearoa/New Zealand and Australia; in both places non-indigenous conservationists see native as good and introduced as bad. The New Zealand National Parks Act requires eradication of introduced species, although, as Kirikiri and Nugent note, there are exceptions. Valued introduced species such as trout and wild horses are in practice excluded. For many Maori, on the other hand, a more pragmatic and less clear-cut distinction is made. Useful introduced animals such as wild pig are accepted, if they are perceived not to have adverse effects on native biota (Kirikiri and Nugent, 1995: 58).

In central Australia many introduced animals are recognized by Aboriginal people as having been incorporated into the ecosystem. In some cases they are also now covered by Aboriginal law: 'Wild pussy cat from here, some rabbit from here too. Pussy cat got Dreaming, some wild pussycat got Law' (quoted in Rose, 1995: 122). In such cases, or where feral animals have become important food resources, Aboriginal people will often oppose eradication programmes.

> The effects of feral animals on the country are not seen as a cause for concern. It is seen as a natural phenomenon that animals eat the grass and raise a bit of dust. To separate the impact of feral animals from native species on these grounds is not seen as logical. People see the contemporary ecosystem as an integrated whole so they don't see some species as belonging while others do not.

TABLE 8.1 Examples of differences in Aṉangu and scientific environmental understandings

Issue	Differences
Taxonomy	• Aṉangu group some that scientists identify as separate species, and vice versa. • Aṉangu provided names of animals not caught in the scientific survey.
Distribution	• Aṉangu increased the species list by recording signs and tracks of animals. • Aṉangu identified 23 spp. restricted to spinifex, with at least another 13 using but not restricted to. Scientists identified 27 spp. as restricted and 15 using but not restricted. Only 12 spp. jointly recognized as restricted.
Ecosystem dynamics	Scientific survey carried out over three wet years. Aṉangu could document impact of dry years, e.g. • Tarkawara (spinifex hopping mice) caching spinifex seed (unrecorded in scientific literature) • Identified refuge areas.
Life history	Aṉangu documented parental care in some reptiles (unrecorded in scientific literature)

Source: Baker and Muṯitjulu Community (1992)

In many areas feral animals are looked on as a resource of the country. Their presence confirms that the land is productive and people derive pleasure from seeing them in the wild. (Rose, 1995: 128)

This is not a simple situation, and discussion of it should not be divorced from an awareness that, because of alienation from land and its radical alteration since European settlement, subsistence choices for most Aboriginal people are extremely limited. Nevertheless, Rose emphasizes, there are significant differences between Aboriginal and European under-standings of conservation management.

Unlike European land managers, who seek to understand and manage their environment, Aboriginal people base their management on practices which have evolved through inter-action. Aboriginal people see themselves as being an integral part of the environment and its dynamics rather than seeking to manipulate the natural world from outside. (Rose, 1995: 90)

In another example, Povinelli records the amusement and frustration of Belyuen women who are continually being asked by researchers to show them 'traditional' foods; 'All of these bush foods are Aboriginal, mangoes and every-thing, animals too, cows; let the researchers photograph anything now' (1993: 128). There is a parallel view among ecologists aware of the profundity of post-European landscape change, particularly in arid areas of Australia. 'There are no "natural" arid systems. Man will always have the responsibility because man, both black and white, has altered the land by his activities' (Graetz, 1988: 138).

These issues relate back to questions of restoration discussed in Chapter 7. The conver-gence of Aboriginal and ecological views about thresholds of change in certain parts of Australia is just one example of the profound rethinking going on. Indigenous knowledge has the potential to make a significant contri-bution to areas of identified need in biodiversity research – for example, the need to

- examine diversity at scales other than species, i.e. genetic diversity, functional type diversity, and, at larger scales, land-scape diversity to understand the interactions within and between these

levels of organization, the effects on trophic pathways, and the consequences for ecosystem functioning;
- understand the resource requirements and population dynamics of species most responsible for ecosystem functioning, i.e. keystone and dominant species, especially in fragmented landscapes. (Sala *et al.*, 1999: 305)

8.2.2 Tradition and change, localness and globalization

Indigenous ecological knowledge is often valued for its traditionality and its relationship to specific localities; it is seen as detailed, place specific and built up over very long periods of time. Questions arise, then, about the resilience of such systems of environmental management in a context of rapid change and globalization. For example, in discussing *hema*, local tradi-tional forms of maintaining sustainability in Saudi Arabia, Saleh (1997) has queried the extent to which they can be maintained in more centralized political systems. As in other examples in this chapter, much of the sustain-ability at issue is cultural in nature.

The idea that indigenous environmental knowledge is necessarily local and at a disad-vantage against the 'whole Earth' perspective that Western scientific conservation claims for itself is challenged by some indigenous leaders.

If we admit that Aboriginal people are fully sentient and intellectual beings, we can admit that they would engage with the effects of the global economy and information society, and that they would bring to these problems interesting and innovative approaches. Sustainable use of natural resources presents options for evolving Aboriginal approaches to the stewardship of their estates. Such approaches, so far, have been based, in part, on the notion that commercial valuation of wildlife constitutes a fundamental protective measure for sustaining populations of species under threat by human impacts. The valuation itself accords the species a status as a potentially non-renewable resource; it must be sustainably managed to enable reproduction. Thus Aboriginal people have an economic preference for involvement in small-scale harvesting ventures, including 'killer' herds of cattle, croc-odile egg harvesting, buffalo harvesting and bio-prospecting. (Langton, 1998: 29)

An example is provided by the Bawinanga Aboriginal Corporation's project to harvest saltwater crocodile eggs, which began in order to stop non-Aboriginal attempts to open up the resource on Aboriginal land (Langton, 1998). Local people gather the eggs from a number of river systems in central Arnhem Land. After hatching they are transported to a Darwin agency which sells them to domestic and international markets. Monitoring of the project by scientists and Aboriginal staff demonstrates the sustainability of the harvest (Webb *et al.*, 1996). However,

> In 1997 senior elders of the local clans rejected a proposal for a trial harvest of adult saltwater crocodiles for the skin trade and local subsistence use of meat, reasoning that commercial harvesting of adults ran counter to the great respect accorded in customary beliefs to these creatures. The religious observance of the ancestral crocodile totemic being in ceremonial life is regionally important, uniting all human descendants and their reptilian cohorts in common interest, and considered to be of great consequence. (Langton, 1998: 64–5)

8.2.3 Traditional harvest

The question of 'traditional harvest' is a major area of contention in Aotearoa/New Zealand debates about wildlife utilization. As we saw in Chapter 7 in relation to vegetation, the pre-human past is sufficiently close in time to be a constant point of reference. At particular issue are Maori attempts to re-establish rights to harvest native birds, with strong opposition from conservationists, predominantly Pakeha (New Zealanders of European descent). Although Maori rights were ostensibly guaranteed by the 1840 Treaty of Waitangi, a series of bird conservation laws enacted since 1862 have effectively prevented Maori from exercising them (Kirikiri and Nugent, 1995; Taiepa *et al.*, 1996).

This debate is intertwined in public consciousness with the extinction of the largest and most vulnerable bird species, particularly the moa, shortly after human settlement. (Aotearoa/New Zealand had no terrestrial mammals other than two species of bat, so questions of faunal use are focused on birds,

fish and sea mammals.) 'Approximately 30% of the bird species present before human settlement became extinct before 1800, with most of those extinctions probably occurring in the first few centuries of Maori occupation' (Kirikiri and Nugent, 1995: 54). By the time of European arrival, a complex of regulatory procedures was in place:

> The regulation of harvest was achieved mainly by a combination of tapu (religious restriction) and rahui (temporary ban) imposed and administered by rangatira (tribal chiefs) and tohunga (experts in the lore relating to natural resources). Fear of divine retribution generally ensured near-absolute compliance with tapu and rahui, and provided a highly effective enforcement system. If divine retribution failed, more down-to-earth measures like muru (confiscation of resources) were enacted. (Kirikiri and Nugent, 1995: 56)

While some have equated this system with a Western conservation ethic, Kirikiri and Nugent liken it more to game management, designed as it was to maximize the harvest and its condition. 'Maori relied heavily on a method of preserving foods such as birds and kiore [the Polynesian rat, *Rattus exulans*, introduced to Aotearoa/New Zealand by the first settlers] in their own fat, so most harvests were timed to coincide with peak fatness' (p. 56).

Moller (1996: 93) reports public debates which question Maori spiritual and philosophical commitment to conservation, often with reference to past extinctions. If traditional *kaitiakitanga* (principles of environmental stewardship) failed to prevent those extinctions in the past, many argue, how can it be relied upon today? (For discussion of parallel debates in Australia, *see* Langton, 1998; Head, 2000). This shows how important it is for researchers and commentators carefully to dissect the debates, and the processes which are under discussion.

There are many issues in this debate; we focus here on the cultural differences. Some details of these differences are presented in Box 8.2, and an overview provided here. To Kirikiri and Nugent the Western separatist perspective is seen in the view that nature occurs 'principally in the network of reserves and wild places beyond the developed lands in which native species survive only if human interference is minimal or solely protective' (1995:

Box 8.2 Summary of responses to the New Zealand Conservation Authority's (NZCA) 1994 Discussion Paper, 'Maori customary use of native birds, plants and other traditional materials'

Differences

Issue	Dominant Maori response	Dominant Pakeha response
Overall ethic	Sustainable use	Preservationist
Hunting of native birds	Important part of culture	Highly offensive – total protection necessary
Rongoa/medicinal plants	Of major concern – access and protection needed	Virtually ignored as an issue
Icon species	A number of species important, including seals in the South, and whales for meat in some coastal areas	Kereru/kukupa/pigeon Toroa/albatross Kuaka/godwit Titi/muttonbirds totara
Harvesting	Sustainability is possible	Will lead to 'slaughter' and 'extinction'
Management frameworks	Many iwi/tribes have prepared extensively researched Resource Management Policy statements setting out appropriate kawa/protocols. Distrust of government and bureaucracy. Criticism of boundaries between land, coast & sea	Legislation and Crown control necessary. Maori seen as lacking in skills, knowledge and commitment
Treaty of Waitangi	Guarantees customary use and participation in management and decision-making	Treaty is 150 years old – Maori should move into modern world
Traditional foods	Have cultural significance over and above food value	Why is traditional harvest necessary when many other foods available?
Pakeha impact on NZ	Pakeha now demanding that Maori bear the sacrifice of 150 years of despoliation	Conservation as part of NZ heritage – 'clean green paradise'

<table>
<tr></tr>
</table>

Common ground between Maori and Pakeha

1. Commitment to fundamental conservation – the demand that nothing more is lost or damaged. Often expressed in terms of quality of experience for future generations.
2. Opposition to uncontrolled harvesting or illegal poaching of native birds and plants.
3. Commitment to participation and demand of the right to be heard and aspirations considered.
4. General acceptance of plant use for weaving and panelling work (but Pakeha concern about use of large timber species for carving).
5. There is a Pakeha tradition of wildlife use, based on a European hunting culture. The traditional rights of duck-shooters and trout fishermen are defended.
6. Need for research and funding.

Source: Cooper (1995). Based on 380 written submissions and a series of consultative meetings

58). That national park system has become a powerful symbol of national identity for the colonial Pakeha society, which 'has found a collective national identity from New Zealand's unique and internationally renowned biota and wild places' (Taiepa *et al.*, 1996: 6). An interesting exception to the Pakeha preservationist approach is the management of marine and freshwater organisms, where sustainability of harvest, rather than non-utilization, is the issue (Moller, 1996: 94).

For Maori, the relationship with native flora and fauna traditionally was and still is far less separatist; this is not just a reflection of their longer history in New Zealand, but also as a reflection of their world view of humans as part of the natural environment ... the very right to harvest placed on them a duty to care for and protect natural resources from 'within' the ecosystem. (Kirikiri and Nugent, 1995: 58; *see also* Moller, 1996)

In practice, a truly bicultural approach to resource management issues will have to acknowledge these differences and find common ground between them (Cooper, 1995). The Rakiura/Stewart Island titi harvest and associated research provides one example (Box 8.3).

Box 8.3 The Rakiura/Stewart Island titi harvest

Titi (sooty shearwater) are one of the most abundant southern-hemisphere birds. Like other shearwaters they are trans-equatorial migrants, over-wintering in the North Pacific and breeding on southern coasts in summer. The Rakiura/Stewart Island (Figure 8.1) titi harvest is the last full-scale bird harvest left substantively in Maori control. (Another species, the short-tailed shearwater, *Puffinus tenuirostris*, has long been important to Aboriginal people in the Bass Strait islands of northern Tasmania (Skira, 1996).)

People leave their mainland homes on 15 March each year to travel to their respective islands to take titi, a process that has both cultural and economic value. According to muttonbirders Margaret Bragg and Jane Davis,

Many descendants of those tupuna (ancestors) walk those same manu (birding grounds) as their people did before them. Rakiura Maori speak of 'ahi kaa' – that is, they have kept their fires burning for succeeding generations. Each year Rakiura Maori return to their turangawaewae, their place of their ancestors, their place to stand on ancestral land. (Taiepa *et al.*, 1996: 14)

The titi harvest is governed by a strict Maori code of regulations, sanctioned by the New Zealand government. However, it is still the subject of criticism from Pakeha environmental groups. Mindful of these criticisms, Rakiura Maori have established a joint research project with a University of Otago team. The aims of the research are to:

- measure the current level of titi harvest and advise on whether it is sustainable (to avoid overexploitation);
- estimate maximum sustainable yield of titi;
- determine what sets the limit of the present titi harvest levels (so that impacts of any future changes to technologies or practices can be predicted);
- determine the diet of titi (so that future studies on their food species can identify threats to titi);
- begin research on impacts other than harvest on titi (e.g. climate change, bycatch, pollution);

- record and compare the understanding of titi ecology, harvest impacts and management practices generated from traditional Maori environmental knowledge and *Kaitiakitanga* with that from Western ecological science and wildlife management.

Setting up this sort of collaborative research involves learning and trust on both sides. It takes much longer, and is much more labour-intensive, than setting up a 'pure' scientific study. About 90 per cent of the discussion centred on the social aims and methods to be employed, and 10 per cent on the science itself. A formal 'cultural safety' contract was drawn up to protect Rakiura Maori intellectual property rights related to their traditional ecological knowledge. The scientific data are jointly owned by them and the University of Otago, and must be published, no matter what the results show.

Source: Summarized from Moller (1996) and Taiepa *et al.* (1996)

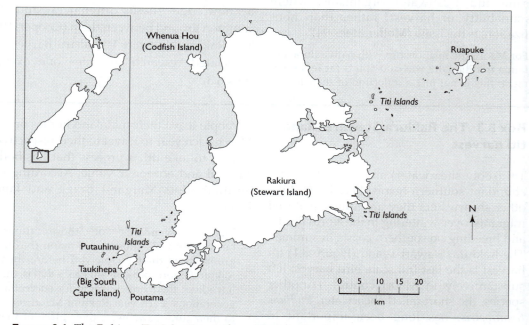

FIGURE 8.1 The Rakiura Titi Islands, southern New Zealand
Source: Taiepa *et al.* (1996: figure 2)

8.2.4 Overlapping landscapes: landscapes of fire

For many Aboriginal people in northern Australia the use of fire is a fundamental aspect of fulfilling responsibilities to their land. While

ecologists and palaeoecologists have focused on the long-term ecological impacts of such fires, the rationale for the practitioners is in the first instance social. Aboriginal people speak of using fire for 'cleaning', 'caring for' and 'looking after' country. (Unburnt country is often understood as 'dirty' or 'wild' country.) Thus a human signature is imprinted on the landscape, and the wild is brought into the human sphere.

In contrast, fire seared itself into the consciousness of many early white Australians as the enemy, a force of destruction. This cultural understanding of fire survives in the hostility of tourists to burnt national parks, in slogans like 'Western Australia … too lovely to litter, too lovely to burn', and in the heroic comparisons which are made each summer between Australia's firefighters and its war dead. There has been a similar understanding in those strands of ecological thought which characterize fire as disturbance rather than as a natural part of ecosystems.

Legislation relating to fire management in Australia's Northern Territory acknowledges only the second of these perceptions of fire, mentioning only the need to 'prevent' and 'control' fire, and nothing of the possibility of using fire actively in the landscape (Hughes, 1995; Head and Hughes, 1996; Illustrations 8.1a and 8.1b). It is likely, however, that Aboriginal rights and obligations to burn country are protected by the more recent Native Title Act;

ILLUSTRATION 8.1 (a) Biddy Simon and Polly Wandanga dig for goanna in burnt pandanus woodland, Keep River area, Northern Territory, Australia.
(b) In contrast, the Bushfires Council of the Northern Territory proclaims, 'We like our lizards frilled not grilled' road sign, Victoria Highway

this is just one of the environmental management issues which are being worked through in the context of debates over native title.

8.3 Developing countries

There are several important themes in the literature on the human landscapes of change in developing countries. These have both similarities and differences with the discussion on indigenous environmental management:

- The relationship between social and environmental sustainability (e.g. O'Riordan, 1998);
- This reiterates the point already made that sociocultural thresholds can be just as important as ecological ones for environmental outcomes;
- A number of case studies demonstrate that the support and involvement of local peoples are crucial determinants of the success of conservation projects (Zimmerer and Young, 1998).

The pace of change is again identified as a crucial issue. In the Andean context Young (1998: 91) distinguishes between 'old' and 'new' frontiers of land use:

> Long-term impacts, accumulated over centuries, have transformed many Andean landscapes, and other long-settled regions, in ways that make them useful to subsistence agriculturalists. Very different are the new frontiers, where current economic processes and road construction are driving deforestation and other land-cover modifications over decadal time periods. ... Prompt action is required to protect intact landcapes from deforestation and to reconnect them regionally. In contrast, deforestation is not the critical issue on old frontiers. Instead, the sustainable use of forest products and the implementation of restoration practices are more appropriate goals.

8.3.1 Adjacent landscapes: protected areas and agricultural land

Human–wildlife conflict, particularly at the interface between agricultural land and wildlife habitat, is seen as a major conservation problem in Africa, reducing local support for conservation efforts. A systematic study around Kibale National Park, Uganda, showed that fields within 500 m of the forest boundary lost an average of 4–7 per cent of crops per season (Naughton-Treves, 1998). The distribution of damage was variable (Table 8.2); elephants, for example, inflicted catastrophic but rare losses. A particularly interesting finding was that damage caused by livestock is comparable to or exceeds that caused by wildlife. However, farmers rarely complain about livestock damage, because 'livestock are providing people benefits and ... communities have established rules of restitution for livestock damage. Under customary law, the owner of a goat or cow must compensate the victim if his or her animals cause crop damage' (Naughton-Treves, 1998: 164). Government, as 'owners' of wildlife and the park, are seen as bad neighbours because they do not offer compensation. As Naughton-Treves points out:

> From a national or international perspective, a loss of 4–7% of planted fields within 500 m of Kibale's boundary appears a trivial price for maintaining threatened habitat and biological diversity. Some would argue that the zone of heaviest crop loss (<200 m from the forest boundary) provides about 3000 ha of 'extra' habitat for wildlife at Kibale. But living within this 'extra' habitat are approximately 4000 farmers who are frustrated and occasionally overwhelmed by crop loss to wildlife and who cannot legally utilize their full range of traditional defensive strategies, such as hunting. Moreover, estimates of average losses mask the great variation in amounts lost by different farmers and villages. To the farmer who has lost an entire year's production in a single night to elephants, average losses are meaningless. (1998: 165–6)

Attitudes to livestock, however, provide a lesson in terms of building local tolerance for wildlife. Farmers need to benefit directly from wildlife conservation, for example through access to game meat, legalized hunting of bushpigs in fields, tourism revenue and employment opportunities.

Similar issues apply in Namibia, where conversion of wildlife habitat to agricultural land is seen as a major cause of biodiversity loss (Richardson, 1998). The returns to wildlife utilization can be shown to be greater than

TABLE 8.2 Amount and distribution of crop damage by animals around Kibale National Park

Animal	Scientific name	No. of events	Percentage relative to total area	Percentage of farms (n = 97)
Redtail monkey	Cercopithecus ascanius	1252	15	88
Livestock	Capra sp., Bos sp.	414	8	79
Olive baboon	Papio cynocephalus	228	24	72
Bushpig	Potamocherus procus	208	15	72
Palm civet	Nandinia binotata	38	1	26
Chimpanzee	Pan troglodytes	146	7	15
Elephant	Loxodonta africana	34	21	8
Black and white colobus	Colobus guereza	11	<1	8
African civet	Civettictis civetta	12	<1	4
Crested porcupine	Hystrix africae-australis	6	<1	3
Vervet monkey	Cercopithecus aethiops	103	2	2
Red duiker	Cephalophus spp.	6	<1	2
Bushback	Tragelaphus scriptus	3	<1	1

Source: After Naughton-Treves (1998: Table 1), simplified

those for livestock (Table 8.3). However, there are a range of reasons why wildlife utilization is not more popular with communal farmers:

- Game farming requires different management skills and more infrastructure than livestock.
- The returns to wildlife tourism do not accrue exclusively to the community, but also benefit private companies and the state.
- People are likely to stick with livestock farming while it is significantly subsidized by the state, e.g. through provision of waterpoints and veterinary services (Richardson, 1998: 555–6).

This illustrates further the complex interaction of processes at different scales. As Richardson argues,

'the problem of global biodiversity conservation becomes one of developing mechanisms to compensate developing countries for conserving diversity. ... If the international community wishes to preserve biological diversity in Namibia, it must pay for it, as development efforts will continue to put pressure on a dwindling global supply of wildlife habitats and biodiversity. (1998: 557)

8.3.2 Sacred groves

Different issues emerge when we consider human activity in relation to plant conser-

TABLE 8.3 Estimated returns to livestock and wildlife utilization enterprises on communal land in Caprivi, Namibia, 1993

Financial return (N$ per annum)	Livestock	Wildlife
Net revenue	2,753,486	3,568,545
Net revenue		
Per ha	1.41	1.83
Per kg	0.10	0.41
Per household	384	498
Net revenue (excl. subsidies)	556,369	3,111,795

Source: After Richardson (1998: Table 4), simplified
Note: Livestock returns come from slaughter for meat sale and from hiring out of draught power. Wildlife returns arise from a combination of photo-tourism, trophy hunting, cropping and live sale

vation in the 'sacred groves' of West Africa. Yet, as with the case of wildlife, perhaps the crucial issue is the interplay between international and national protective mechanisms and local traditions and aspirations. Sacred groves are forest fragments varying in size from single trees to hundreds of hectares, protected from excessive disturbance by a variety of social and religious mechanisms. There are about 2000 such groves in Ghana, and they are also known from India and north-eastern Thailand (Decher, 1997). In Chapter 6 I made reference to sacred groves as part of World Heritage cultural landscapes in the Philippines and Zimbabwe. The term 'sacred grove' glosses over a great deal of social and ecological variability; we concentrate here on some of the common issues.

According to Decher (1997: 1011), 'the ecological value of sacred groves rests on the fact that most of them have been more or less undisturbed for up to several hundred years'. He reports on (unresolved) debate over whether the groves 'are remnants of the original climax vegetation' (p. 1011); as we saw in Chapter 3, these debates are enormously problematic in the context of tropical rainforest and wet-dry forests. As Decher's own study shows, lack of disturbance does not equal lack of use. In Ghana, traditional religion, centred on a deity called Kpalevorgu, has been effective in protection of the groves from fuelwood cutting. However, they are used as a source of

hardwoods, fruits, medicinal plants and small game (Figure 8.2). In India, sacred groves have been found to contain threatened plant species. They are also sources of rattan (*Calamus* sp.), considered a sacred plant by some groups, and of fuelwood and valuable timber (Decher, 1997: 1012). All such patches provide important small mammal habitat.

In a context of population growth and land-use pressure, it is not hard to see the ecological threats to patchy forests: 'over-exploitation of resources, clear-cutting, bush fires, and the introduction of exotic plants' (Decher, 1997: 1016). Yet what is most striking about the evidence assembled by Decher is that the most relevant aspects of resilience and fragility to be discussing are the intricate networks of socio-cultural process by which the groves are currently maintained. These make a mockery of any simple designation of legal or heritage protection. In India, for example, breaking of the taboo by even moderate Forest Department use would lead to poaching of wood by villagers. Writers on the Ghanaian situation emphasize that with the gradual breakdown of traditional religion, survival of the groves depends on local people developing non-religious incentives. Incorporation of groves into the national reserve system could be counter-productive if it takes away local jurisdiction.

The idea of 'sacred groves' is an iconic one for Western conservation interests, with both forest fragments and local people providing

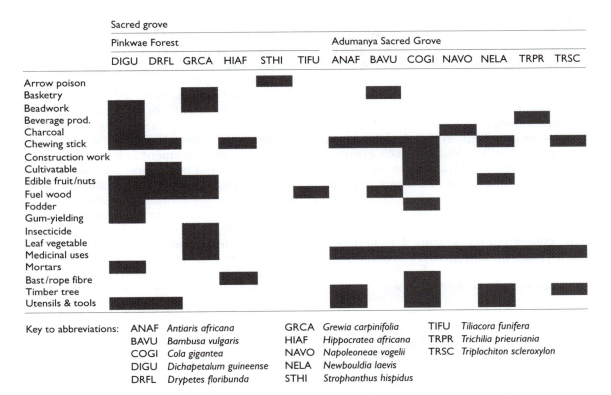

FIGURE 8.2 Some uses of common tree and climber species found in sacred groves on the Accra Plains, Ghana
Source: After Decher (1997: table 1). By permission of Kluwer Academic Publishers

images of the vulnerability of biodiversity before the advancing wave of development. While there is truth in this imagery, it is important not to obscure the specific and variable processes involved. The sacred grove is not an arena of human absence; on the contrary, active human intervention has been crucial to its maintenance. The most significant 'thresholds of irreversibility' in this instance may be cultural rather than ecological.

8.3.3 Contesting the iconography of climax forests

Sagarmatha National Park (SNP), Nepal, on the southern flanks of Mt Everest (Figure 8.3), also has iconic status within international conservation. Created at a time (1976) when the International Union for the Conservation of Nature (IUCN) criteria specified a national

park to be 'not materially altered by human exploitation and occupation … [and] where the highest competent authority … has taken steps to prevent or eliminate as soon as possible exploitation in the whole area' (Brower and Dennis, 1998: 187), SNP has long been home to the Sherpa people. Management debates in recent decades have revolved around a particular dimension of human activity, its impact on the region's forest landscape. Brower and Dennis argue that 'arbor-centric management persists despite a growing uncertainty about the processes that shape the area's forest landscape … and in the face of continuing local resentment about the infringement of accustomed access to resources' (p. 185).

Thinking about human impacts on the forests of the Khumbu area where SNP is located is tied into wider debates about original forest states that are themselves heavily influenced by Clementsian views of succession. Like many of the sacred groves, Khumbu's forests have often been assumed to be degraded remnants of something else, in this case of an uninterrupted fir-dominated (*Abies spectabilis*) forest landscape below 4100 m. In this scenario the arrival of Sherpa settlers some 25 generations ago initiated deforestation through tree felling and livestock grazing, a process accelerated through increasing tourist pressures in recent years. However, in a study of forest plots around the most intensely settled area of Kunde, Khumjung and Namche (Figure 8.2), Brower and Dennis (1998: 195) showed that 'contemporary grazing has at most a limited effect on the regeneration of forest; it is significant – if at all – only when acting in conjunction with other site stresses'.

While the stands showed variable characteristics, there is evidence of forest cover increasing in the 3400–3800 m zone. Sparse stands and lone trees 'are pioneers, not remnants of formerly more dense forests' (Brower and Dennis, 1998: 202). They suggest that establishment of fir seedlings in shrub/grassland is a rare event because the seedlings are vulnerable to desiccation. Once past the establishment phase, however, shrub/grassland sites are suitable for fir growth. *Abies* is relatively unpalatable to large herbivores, and tolerates moderate browsing when healthy. Another implication of the arbor-centric perspective is that it is prone to overlook the richness and diversity of the *Rhododendron lepidotum–Cotoneaster microphyllus* shrub/grassland, often characterized as 'degraded' (p. 202). Brower and Dennis further

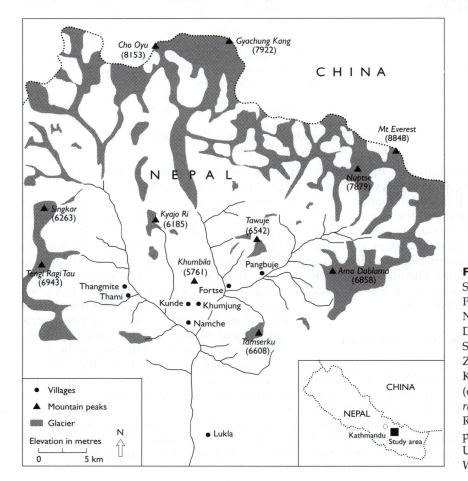

FIGURE 8.3
Sagarmatha National Park in Khumbu, Nepal (Brower and Dennis, 1998) Source: Karl S. Zimmerer and Kenneth R. Young (eds), *Nature's geography.* © 1998. Reprinted by permission of The University of Wisconsin Press

speculate that Sherpa livestock grazing may actually be increasing fir forest extent by reducing fuel load and thus fire frequency.

As with all the case studies presented here, this is not to deny the obvious fact that human activities can be detrimental to ecosystems, whether assessed by biodiversity, sustainability or some other measure. It is to demand more explicit consideration of the rates and directions of human influences within situations of change. Blumler argues that

> although the human imprint on the Near Eastern landscape is major, effects are often conflicting and diffuse rather than unilinear and utterly degradational. For instance, people destroy woody plants when they gather them for fuel or construction material, but may *favor* their establishment and survival when domestic animals graze and weaken the competing herbs, or when soil erosion exposes rocks that then protect woody seedlings and provide them a favorable, summer-moist microclimate. (1998)

As Blumler points out, the Near East has a history of agropastoralist impact as long as anywhere in the world. Most understanding of this impact, he argues, derives from 'the traditional equilibrium view of nature, and associated linear models of vegetation change under human impact' (p. 218). Trees are associated with climax (Figure 3.2), thus tree-planting becomes a high-priority conservation activity, often at the expense of formerly species-rich landscapes.

Blumler and colleagues' work in open oak (*Quercus ithaburensis*) woodland in the Lower Galilee showed among other things that variation in disturbance regime favoured diversity. Elaborating ideas put forward by Connell (1978), they argued that each type of distur-

bance, including anthropogenic ones (fire, drought, ploughing, etc.), favours a particular suite of species (Figure 8.4). 'All species remain in the system as long as the different disturbance types cycle through with sufficient frequency' (Blumler, 1998: 227). This suggests that flexible, varying land use, with changes in impacts, should be encouraged in the Near East and Mediterranean ecosystems.

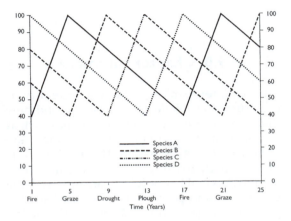

FIGURE 8.4 Blumler's disturbance-heterogeneity hypothesis (1998). If a species is favoured by one type of disturbance but not by others, it will increase in abundance after the former but decrease after the latter. As long as the type of disturbance that it 'needs' cycles through frequently enough, it can remain in the landscape. If there are many species adapted to different sorts of disturbances, then overall species diversity can be greatly enhanced if disturbances are also diverse

Source: Karl S. Zimmerer and Kenneth R. Young (eds), *Nature's geography*. © 1998. Reprinted by permission of The University of Wisconsin Press

9

Identity, heritage and tourism

9.0 Chapter summary

This chapter explores the way identity is constructed in relation to landscape, and how that is expressed in management issues such as heritage and tourism. Two themes are dominant in this discussion. The first is the role of time, particularly past time, in embedding identity. Several temporalities are involved: the passage of time, perceptions of the past and the pace of change in different arenas of practice. The second theme is the contests that emerge from overlapping or competing identities in landscape. The prime example of this is increasing concern with heritage, a theme that echoes issues of preservation and restoration discussed in Chapter 8. The two are connected by two important dimensions of heritage identified by Lowenthal (1996): origins or priority – 'being first', and lineage and kinship – 'being innate'. With increasing tourist interest in cultural landscapes, particularly old ones, these issues play out in terms of present and future management.

9.1 Identity and the past

In Chapter 8 we considered the role of indigenous people and issues in environmental management. Indigenous people often articulate their interest in land and resources with reference to time, to a long period of interaction with the land, and thus an accumulation of

detailed knowledge. This begs the question of how long it takes to develop indigeneity and how it is expressed in different contexts, a question taken up by Dominy's study of (Pakeha) high-country sheep station families in South Island Aotearoa/New Zealand (Dominy, 1995).

The assertion of native status by these families occurs as part of contemporary Aotearoa/New Zealand political discourse, specifically in the context of a land claim before the Waitangi Tribunal. Pakeha New Zealanders, like contemporary descendants of colonizing peoples in Australia, Canada, the USA and South Africa, are in the process of creating a new sense of belonging in the land that involves looking both backwards and forwards. An important element of this is the differentiation from imported and inherited traditions (Dominy, 1995).

In the claim under discussion, 'Ngai Tahu people seek remedies for the Crown's failure to protect rights guaranteed to the tribe by the Treaty of Waitangi' (Dominy, 1995: 361). The land claim included land held under pastoral lease tenure by 360 sheep-farming families of European descent. There are many issues involved; we concentrate here on the way the high-country farmers represented the historical importance of their relationship to the land to the Waitangi Tribunal. Three individuals represented the High Country Committee of Federated Farmers. Hamish Ensor presented evidence that he understood Ngai Tahu's relationship with the pastoral lands, for example trails through the Rakaia valley to the west coast for greenstone. He also asserted his own family's social connection to the land:

> If any group of New Zealanders can claim to be the indigenous people of the pastoral lease land perhaps it is the lessees themselves as they are the only people in the history of New Zealand to have actually settled on and worked the land in question. My people, regardless of race or creed, including members of the Ngai Tahu tribe[,] consider themselves to have the indigenous feeling of the high country. In many cases the occupation by these lessees extends back over four or five generations. (quoted in Dominy, 1995: 363)

Ensor's expression of attachment has both parallels and differences with Ngai Tahu's claim. Both refer to kinship and generational inheritance as evidence of commitment to the land. Ensor's argument in relation to 'settling on and working' the land goes back to colonial ideas that did not recognize the land relationships of non-agricultural peoples.

A second pastoral representative, Iris Scott, emphasized belonging through detailed knowledge of the environment, including the vagaries of local climate and the implications of changing terrain conditions for stock movements. In this she reflected attitudes commonly expressed to Dominy by other farmers, who thus strategically situate their own local knowledge in the context of the national interest: 'The pastoral lease symbolizes a relationship of shared ownership between the runholder and the Crown in order to "preserve the land for the benefit of all New Zealanders"' (Dominy, 1995: 365).

Pastoralist Jim Morris equated his own identity with the land with that of Maori, arguing that his 'feeling for the land and for the people who also live and work there is of an order that the Maori people would understand' (Dominy, 1995: 366). In doing so Morris challenged homogenized and dichotomized views that only Maori can speak of ancestral and spiritual concerns for the land, and that Pakeha farmers think of it only in practical and economic terms. In a series of interviews on the west coast of the South Island, Kirby (1993) also showed that both Maori and Pakeha informants expressed particular attachments to local place.

The creation of identity in a colonizing context is addressed also by Kealhofer (1999) in Virginia, USA, drawing on a combination of historical, archaeological and architectural evidence. This study also challenges simplistic models of identity creation and landscape transformation over time. If identity is a complex and multi-layered thing, then so will be the landscapes in which it is worked out. Creating public landscapes – of social hierarchy or government control – took longer than did creating private ones such as gardens. This results in lags and discrepancies that need to be considered in interpreting changing shape and meaning over time (Kealhofer, 1999).

One relevant aspect of time here is time since colonization. Kealhofer argues that when worlds are still 'new', we can reasonably expect discrepancies between constructed and conceived landscapes. (This contrasts with times and places of long habitation, where the distinction between the two may have become quite blurred.)

> Colonists 'conceived' of their new landscape – often described by English explorers as Eden – but they did not configure it with any complexity or depth of meaning. For example, sacred points of reference did not exist. It takes time to embed meaning, even in a little modified 'conceived' landscape. (Kealhofer, 1999: 59)

Kealhofer (1999) explores these points of contradiction and discrepancy as a means to understand the trajectories of social change. Throughout the seventeenth century houses gradually became more permanent, and space within them became more differentiated. Communal and private space were separated, and servants were moved into separate structures. In contrast, 'gardens increasingly link the house with the larger landscape' (p. 75). Even within gardens there was considerable variability, as excavations of two seventeenth-century elite gardens show. The landscaping of Green Spring plantation shows how Governor Berkeley defined himself as both Virginian and Englishman. Excavations of the garden showed it falling away from the terrace of the house, within a wider landscape of meadows and trees as the forest had been cleared for shifting tobacco cultivation. Kealhofer sees this as linking the plantation into an increasingly domesticated landscape: 'the garden becomes part of the natural world as viewed from the house, and the stature of the house was increased when viewed from afar' (p. 72). Constant exchange of plants, gardeners and ideas within the colony and across the Atlantic formed another network within which landscapes held meaning.

The garden excavated at nearby Bacon's Castle shows a quite different configuration, one that does not link the house to the wider landscape. This garden was more in medieval style, with geometric vegetable and herb beds divided by wide formal paths. Attention in this garden is focused internally. Whereas Governor Berkeley's landscape garden reflected social and political interactions out into and across the countryside, Arthur Allen of Bacon's Castle was more interested in control and order of his own domain. It is suggested that 'Allen was reformulating a more medieval identity that focused primarily on his extended household, and himself as a traditional manor lord' (Kealhofer, 1999: 74).

9.1.1 National identities, landscape and the past

If identity is complex and operates at a range of scales, one of the most powerful levels is that of nation (Illustration 9.1). 'Nations without pasts are a contradiction in terms and archaeology has been one of the principal suppliers of the raw material for constructing pasts in modern struggles for nationhood' (Rowlands, 1994). The association of archaeology with the nation-building project is particularly strong in East Asia (Ikawa-Smith, 1999), from where I draw a number of examples.

Ikawa-Smith (1999) identifies at least three models of the way national origins are constructed in relation to archaeological evidence. One emphasizes indigenous or *in situ* development, for example China. A second, the 'continuity with assimilation' model, incorporates influx of external sources into a basic cultural continuity, as in Japan. Third, there is a 'single ancestral antecedent' model, whereby a pure and homogeneous group is assumed to have moved into an area. In the example of Korea this homogeneity tends to be emphasized because the relevant archaeological remains are now found in three sovereign states (Pak, 1999).

These issues are all cross-cut by the historical experience of colonialism. Areas of Africa, Asia and the Americas which experienced European colonialism were also subject to a distinctive archaeological heritage, in which local cultures were regarded as 'a living museum of the past' (Glover, 1999: 594). These cultures were seen as not being able to change without external stimulus, unless it was to wither and decay from a higher level, as in the case of the ruined towers of the Cham peoples of Vietnam

ILLUSTRATION 9.1 'The heritage and identity contradictions of the former GDR: Marx and Engels stand in front of its Parliament building in the former East Berlin, while behind is the Dom, commissioned by Kaiser William II and opened in 1905. Damaged in the Second World War, the building was restored by the GDR government as part of a policy of appropriation of past German culture to claim legitimacy for itself' (by permission of Graham *et al.* (2000: 67))

(Glover, 1999). Scholars working in these areas in the post-colonial period have had to rethink many issues. Two important ones are the place of indigenous and minority groups in national pasts, and the way colonialism itself is commemorated in the landscape. Indigenous peoples have variously been erased from, incorporated into and appropriated as part of national pasts. Rao (1994) reports the deliberate removal or renaming of British and Japanese colonial monuments respectively in India and Korea. On the other hand, they have been preserved in Mozambique as symbols of the people's struggles for freedom.

These discussions are inevitably interlinked with landscape, and power over land. We start here by looking at England and its former colony Australia, where national identities are entwined with both landscape and the past, but in very different ways. In both places, rural landscape – 'countryside' in England, 'bush' in Australia – has been at the heart of discussions of national identity, discussions led by largely urban populations (Lowenthal, 1996). There are differences in the tenor of debate. For example, class-based differences over rights of access to countryside (Blunden and Curry, 1990) seem peculiarly English, although Aboriginal Australians would find many parallels in their own land rights struggles.

A clearer difference is the way each nation thinks about a landscape past. While both make reference to ancientness and, to varying degrees, stability of landscape, there are differences in how the human presence is considered. In England it is the past of a peopled landscape:

> English stability is notably enshrined in landscapes that bear the stamp, as its champions fondly say, of centuries of countrymen and women – even of surviving aboriginal cattle. An environment chief lauds stewardship that leaves much of rural England 'as she was: changeless in our fast-changing world'. (Lowenthal, 1996: 186)

By providing reassurance that some things remain stable, permanent and enduring, the countryside upholds the *status quo*.

As Australians come to terms with the antiquity of Aboriginal occupation of the land,

a national identity that alluded to the ancientness of the landscape itself is in the process of being recast (Head, 2000). In 1963 it was still possible to think of the dead heart of the country as providing a cultural blank slate, as expressed in the opening lines to Moorehead's *Cooper's Creek*:

> Here perhaps, more than anywhere, humanity had had a chance to make a fresh start. The land was absolutely untouched and unknown, and except for the blacks, the most retarded people on earth, there was no sign of any previous civilization whatever: not a scrap of pottery, not a Chinese coin, not even the vestige of a Portuguese fort. Nothing in this strange country seemed to bear the slightest resemblance to the outside world: it was so primitive, so lacking in greenness, so silent, so old. It was not a measurable man-made antiquity, but an appearance of exhaustion and weariness in the land itself. (Moorehead, 1963: 1)

In 1978 Lowenthal described Australian (and, comparatively, American) attempts to find a usable past when faced with recorded histories of 'embarrassing brevity':

> To compensate for the absence of a long recorded history, we are prone to stretch history back into the mists of pre-history. The Australian heritage incorporates not only the few decades since the European discovery but the long reaches of unrecorded time comprised in Aboriginal life and, before that, in the history of nature itself, the animals and plants and the very rocks of the Australian continent. Thus the felt past expands, enabling Australia to equal the antiquity of any nation …
>
> The Aboriginal past is largely absent from this Australian primordial heritage, partly because Australians and Aborigines are still antagonistic, partly because earlier epochs yield only paltry Aboriginal remains. A few flints and kitchen middens inspire no awe comparable to Mayan pyramids or mid-Western Indian mounds. … Only if Aboriginal 'hallowed ashes' became conspicuous monuments in the Australian landscape could the Aboriginal past form part of the mythic heritage that nature alone now supplies. (Lowenthal, 1978: 86, 88)

Even as he wrote, the middens, flints and hallowed ashes were in the process of becoming such monuments (Illustration 9.2),

ILLUSTRATION 9.2 Walls of China, Mungo National Park, New South Wales, Australia. Archaeological remains eroding from these dunes around now-dry lake beds date the human occupation of Australia and burial rituals to beyond 40,000 years. This area was proclaimed as a World Heritage site in 1981

and understanding of Australia's prehistoric past has been transformed in the intervening decades, as discussed in Chapter 2. By 1993, elements of an archaeological consciousness were being incorporated into official nationalisms. This was expressed by Prime Minister Paul Keating in his November 1993 Address to the Nation on the Native Title Bill.

> The Bill will necessarily be complex, but this evening I want to cut through the complexity to some of its simple principles. First we need to get the background straight. Over tens of thousands of years Aboriginal people had developed a complex culture built on a profound attachment to the land. ... Yet this most remarkable fact about Australia – this oldest continuous civilisation on earth – has until now been denied by Australian law. (Keating, 1993)

Although archaeology is not mentioned, the priority accorded by antiquity is implicit in the Preamble to the Native Title Act 1993:

> The people of Australia intend ... to ensure that Aboriginal peoples and Torres Strait Islanders received the full recognition and status within the Australian nation to which history, their prior rights and interests, and their rich and diverse culture, fully entitle them to aspire. (p. 2).

Aboriginal people have expressed a range of views, both positive and negative, about archaeology and the way it interprets the past. These are not explored here. Rather, I am interested to demonstrate the dilemmas for non-Aboriginal Australia in constructing an identity around a landscape past that belongs essentially to someone else. While past English landscapes may have had great differentials in terms of power and access, that history is one that people can have ownership of at some shared level. It remains to be seen whether this can be the case in Australia. In Australia there is the further issue that some key landscapes in the process of nation-building are on the other side of the world (Illustration 9.3).

England and Australia provide one particular comparative example, but the issue of incorporating pasts, including the archaeological past, into contemporary identity is a widespread one. Some of the most contested landscapes are those where an indigenous minority maintains a past and a present. The traditional lands of the Sámi, or Lapps, lie within four present-day countries: Russia, Norway, Sweden and Finland. To some extent their political struggles vary in those different contexts (e.g. Brantenberg, 1991), but a shared concern is the way in which Sáminess is presented and negotiated. In this example the focus is on how an archaeological past is incorporated into discussions of Sámi identity in Sweden.

The Sámi occupy an area in the interior of northern Sweden where they have been intensive reindeer pastoralists since *c.* AD 1600. The seasonal movements on which this pastoralism depends have become much more difficult in the past two centuries owing to land use pressures and the attitude of the Swedish

ILLUSTRATION 9.3

'The road to nationhood through sacrifice in war: the Australian National Memorial in Villers-Bretonneux Military Cemetery, Somme, France' (by permission of Graham, *et al.* (2000: 194))

state. Many Sámi chose or were forced into new, more sedentary ways of life, consequently losing some aspects of their Sámi identity (Mulk and Bayliss-Smith, 1999). This was exacerbated by state policies which recognized limited grazing and use rights for the dwindling number of reindeer herders, but none for other Sámi. As a consequence, 'Sámi ethnicity came to be regarded as practically synonymous with the practice of reindeer-herding' (Mulk and Bayliss-Smith, 1999: 361). Other subsistence practices such as fishing and farming, engaged in by more Sámi than is reindeer herding, are not seen to have ethnic significance because they are not exclusively Sámi activities (Brantenberg, 1991).

Some of these policies have altered with the recent nomination of Laponia for World Heritage listing as both natural and cultural landscape (*see also* Chapter 6). The significance of the area for the Sámi goes beyond livelihood to sacred and symbolic meanings, expressed in 'rock carvings, Dreaming spots, landmarks that feature in myths, sacred mountains and sacrificial sites' (Mulk and Bayliss-Smith, 1999: 367). The Swedish government submission to UNESCO had to recognize and protect Sámi rights as part of the case for a World Heritage cultural landscape.

In this political process the archaeological past of the Sámi has been used in contradictory ways, as elaborated by Mulk and Bayliss-Smith (1999). Until well into the twentieth-century archaeology reinforced the hegemonic relationship between majority groups in Scandinavia (Norwegians, Swedes) and minorities like the Sámi, mainly by rendering the latter invisible. For example, the museum of Vuollerim 6000 år (Vuollerim 6000 years ago) is a popular tourist destination along the road to the 'wilderness' landscapes of the Laponian World Heritage Area in northern Sweden. The archaeology of the 6000-year-old settlement is interpreted as a winter base camp for a hunter-gatherer group, with no direct evidence of ethnicity. Displays and publicity material (Figure 9.1), however, portray these people as 'tall, blond, blue-eyed people with stereotypical Germanic features' (Mulk and Bayliss-Smith, 1999: 380). This supports old nationalist myths that portray Norway and Sweden as ancient cultures with Stone Age roots, and Sámi as latecomers from an Asian homeland. Mulk and Bayliss-Smith stress the intimate links between the acquisition of ethnographic and archaeological knowledge of the past, and the colonial process itself:

> The collection of artefacts and skeletons, old and new, and the recording of curiosities such as myths and vocabularies, was in a real sense the scientific parallel of the economic and territorial conquest that incorporated these lands and peoples of the north into the emerging nation states of Norway, Sweden and Russia. Such knowledge was constructed and represented in ways that served to emphasise the contrast between the societies of the 'civilised' centre and those of the 'primitive' periphery. (1999: 379)

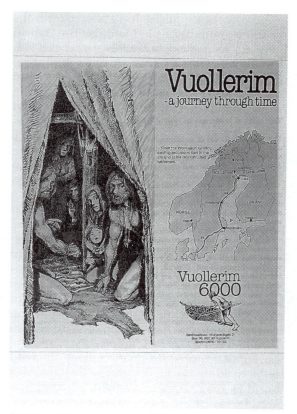

FIGURE 9.1 'Vuollerim 6000 år: not-so-hidden messages about ethnicity' (source: Mulk and Bayliss-Smith 1999: figure 24.6). By permission of Taylor & Francis Books Ltd

The Sámi situation provides just one example of the well-documented link between colonialism and archaeology (e.g. Thomas, 1994; Head, 2000). The opposite tendency, the romanticizing of the past, is also not confined to the Sámi. It often goes hand in hand with the depiction of indigenous peoples as the 'original conservationists'. This example relates to the interpretation of so-called 'stallo' sites: groups of former huts or dwellings that are defined by central hearths and surrounded by oval-shaped banks of soil. The sites date to the late Iron Age and Nordic medieval period. For nearly a century there has been debate about their function and their connection or otherwise to the Sámi. While there are still areas of disagreement, most archaeologists now agree that

the stallo dwellings are of Sámi origin, and are the remains of tent structures like the *kåta* of the recent past. There is also agreement that these high-altitude settlements can have been occupied only in the summer or autumn, and that they were used by people who had control of domesticated reindeer as draught animals. (Mulk and Bayliss-Smith, 1999: 384)

What is under dispute is why the Sámi were occupying sites in the high mountain areas. At particular issue is the question of whether they were hunting for trade in fur and meat, or whether these sites indicate the onset of full reindeer pastoralism, comparable to more recent Sámi lifeways, about 1000 years earlier than the usually accepted date. The latter proposition, advocated by Storli (1996), constitutes for Mulk and Bayliss-Smith 'the past-as-wished-for'. They argue instead that extensive wild reindeer hunting, as evidenced by widespread hunting pits dated *c.* AD 500–1350, could not coexist with domestic herd management, and thus that pastoralism had not yet become established. Our task here is not to try to resolve any particular archaeological debate, but to understand how various interpretations of the past are incorporated in complex and contested ways into identity politics. Mulk and Bayliss-Smith summarize this well:

The lingering assumptions of Social Darwinism and cultural evolution still mean that for some Sámis, as for many other people in the world, a prehistory of pastoralism seems more dignified than one dominated by hunting and gathering. If the stallo sites were to be seen as a cultural landscape of early reindeer pastoralism, then Sámis who had adopted the hegemonic view of cultural evolution would feel that their cultural identity today had been significantly strengthened. Grazing rights for modern reindeer-herding might also be seen as having a stronger basis if a millennium of continuous pastoralist land-use could be established.

In other ways, too, modern Sámi cultural identity would be confirmed if we were to interpret cultural landscapes in ways that suggest that Sámi pastoral society had very ancient roots. Scandinavian society today focuses substantially, perhaps disproportionately, on reindeer-herding as a symbol of Sámi cultural identity. This is a situation in part imposed by the nation states that have equated Sámi ethnicity with reindeer pastoralism, but in part it is created by the Sámi themselves, as a nostalgic reaction to the abandonment of reindeer-herding by the majority of the Sámi people. (1999: 389)

The value of a hunter-gatherer past changes according to context. Once ignored or derided as evidence of a lower state of civilization, it is now more in tune with contemporary environmental sensibilities. (If it can take us back to human ancestry itself, as in the case of China, it is in another realm of continuity altogether (Ikawa-Smith, 1999)). However, the issue is also related to whether the contemporary nation can 'own' this part of the past, as the examples above show. In Japan a shift in thinking is beginning to incorporate Jomon hunter-gatherer archaeology into Japanese national identity (Habu and Fawcett, 1999). Particularly important in this process has been public interest in Sannai Maruyama site (3500–2000 BC) in Aomori Prefecture, northern Honshu (Figure 9.2 and Illustrations 9.4a, 9.4b and 9.4c).

Excavations at Sannai Maruyama have revealed 'over 7 pit-dwellings, approximately 20 long houses, about 100 remains of raised-floor buildings, approximately 250 adult grave pits and 800 burial jars for infants or children', several large middens and mounds, and an enormous range of utilitarian and artistic artefacts (Habu and Fawcett, 1999: 587–8). These are thus very respectable hunter-gatherers to have in the national family tree, a shift from their previous depiction as 'barbarous wanderers' (p. 592). (Indeed, Jomon archaeology has been central in international archaeological debates over the variability and complexity in hunter-gatherer societies.) Nevertheless, the shift in Japanese thinking from a situation where Japanese cultural origins have always been traced to the rice-producing and socially stratified Yayoi (*c.* 300 BC–AD 300) and Kofun periods is seen by Habu and Fawcett as a significant one. Nationalist focus after the Second World War linked Japan as a rice-farming nation to an ancient agrarian past.

Several important factors have contributed to public interest in Jomon archaeology (Habu and Fawcett, 1999). Archaeologists working at Sannai Maruyama have been active in disseminating excavation results to the public, assisted by intense media interest. The people of Aomori Prefecture have developed intense regional pride in the site, and the prefectural government has been keen to develop it as a tourist attraction. In 1998 the governor declared a focus on cultural tourism, pointing out that Sannai Maruyama 'added temporal depth to Aomori's natural beauty and traditional cultural assets'. He further called Aomori 'the hometown of the Japanese nation' and then linked it symbolically to international society (Habu and Fawcett, 1999: 591).

On the one hand, pushing Japanese ethnicity further into the prehistoric past can be seen as a conservative reaction which reifies traditional ideas of cultural and biological homogeneity. On the other hand, argue Habu and Fawcett (1999), increased interest in Jomon sites opens the way to a re-examination of the whole question of Japanese ethnicity, and for recognition of prehistoric diversity. This might 'provide space for a modern-day acceptance of ethnic diversity within Japan' (p. 592), particularly for the marginalized Ainu and Ryukyuan people of Hokkaido and Okinawa.

In India a traditional past was sought as a bulwark against the disarray which marked the end of British colonialism. This has resulted in a struggle between Hindu and Muslim pasts, epitomized in the struggle over the site of a mosque in the northern city of Ayodhya (Rao,

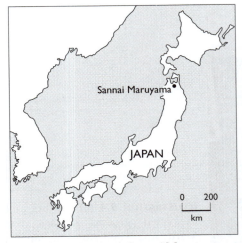

FIGURE 9.2 Location map, Sannai Maruyama

ILLUSTRATION 9.4 (a) Remains of circular stone monument, Komakino, near Aomori, Japan. It was built about 4000 years ago during the Jomon era. In 1995 the site was designated as a national historical treasure (by permission of Mary-Jane Mountain)

ILLUSTRATION 9.4 (b) Sannai Maruyama site, Aomori, Japan. The tower and huts are reconstructions based on the archaeological excavations (by permission of Mary-Jane Mountain)

ILLUSTRATION 9.4 (c) Visitors look at excavated ground surface, Sannai Maruyama site, Aomori, Japan (by permission of Mary-Jane Mountain)

1994). Majority Hindu groups claim that a sixteenth-century mosque, Babri Masjid, was erected on the site of an ancient Hindu temple to the god/king Rama. According to the *Ramayana*, one of the Indian epics, Rama was born in and ruled from Ayodhya (Rao, 1994: 156). Textual, historical and archaeological evidence have all been invoked in these contested pasts (Golson, 1995), but it became more than an academic argument when the mosque was torn down by Hindu fundamentalists in December 1992. It is estimated that more than 1000 people died in the wave of violence which followed (Colley, 1995).

The dispute has moved beyond a debate over archaeological or historical 'facts', with the site increasing in importance to both Hindus and Muslims as a result of the controversy. One consequence, as Rao observes, is that in the struggle for symbolic control of space, previous ambiguities become more reified. 'Ayodhya has never really occupied the absolute position in the Hindu mind that Mecca, for instance, does in the Muslim. Liminality and ambiguity are a part of the Hindu Weltanschauung' (Rao, 1994: 163). By liminality, Rao refers not to physically marginal space, but a sort of magical space which is 'betwixt and between'. For example, each year the story of Rama is played out in the town of Ramnagar over a thirty-day period. Different parts of the town 'become' different aspects of the story, including Ayodhya. For this month 'Ramnagar and its inhabitants enter a symbolic, liminal space' (159).

9.2 Contested identities: landscapes of heritage

All the debates above allude to the concept of heritage, so it is important to tease out some of the assumptions buried, and not-so-buried, in the valuation of the past that is implied by this term. Lowenthal's (1996) analysis of heritage helps us here. Importantly, the themes of *stability, origins* and *lineage* in his work resonate with themes we have already discussed in terms of ecological protection (especially in Chapters 3 and 7). Once again it is instructive

for us to compare discussions in the natural and cultural management fields.

Lowenthal starts by exploring the distinction between history and heritage. Understanding this distinction is crucial to understanding the partisan nature of heritage:

> History tells all who will listen what has happened and how things came to be as they are. Heritage passes on exclusive myths of origin and continuance, endowing a select group with prestige and common purpose. ... the very notion of a universal legacy is self-contradictory ... confining possession to some while excluding others is the raison d'être of heritage. ...
>
> Heritage builds collective pride and purpose, but in so doing stresses distinctions between good guys (us) and bad guys (them). (1996: 128, 230, 248)

This distinction, which Lowenthal discusses mostly in terms that we would refer to as 'cultural' heritage, stimulates us to think also about the processes of invoking 'natural' heritage. The relationship between history and heritage is also paralleled by that between geography/ecology and natural heritage. Why are some environments and ecosystems valued and others not, for example? And who does the valuing?

Lowenthal identifies two different strategies of heritage-making. One collapses 'the entire past into a single frame' and makes it different from the present. This can be for better or worse; the good old days, or the bad old days. A second strategy stresses the likeness of past and present. '[C]oalescing past with present creates a living heritage that is relevant because it highlights ancestral traits and values felt to accord with our own' (1996: 139). Several examples presented above accord with this strategy. Heritage alters history in at least one of three ways: it updates it; it highlights and enhances aspects now felt admirable; and/or it expunges shameful or harmful aspects of the past (p. 153).

9.2.1 Stability, origins and lineage

The desire to preserve heritage, like concern over rainforest loss, is intensified by the perception of accelerating change (Illustration

9.5). For Lowenthal, this helps to explain our focus on stability, however artificial:

> Beleaguered by loss and change, we keep our bearing only by clinging to remnants of stability. Hence preservers' aversion to letting anything go, postmodern manias for period styles, cults of prehistory at megalithic sites. Mourning past neglect, we cherish islands of security in seas of change. … Legacies at risk are cherished for their very fragility. The heritage of rural life is exalted because everywhere at risk, if not already lost.
>
> Fending off irreversible change, we preserve, restore, or replicate. Any extinction, even of pestilential germs, becomes a crime against the legacy of diversity. (Lowenthal, 1996: 6, 11)

If stability rather than change is valued, it is a short step to the importance of being first. Issues of priority, precedence, primordiality and antiquity are central to heritage questions, as we saw in the discussion above. Precedence

ILLUSTRATION 9.5 Juxtaposition of old and new, Seoul, South Korea (by permission of Gordon Waitt)

can be transferred along appropriate lines of lineage and kinship; related ideas reappear in more metaphorical guise in discussions of race, blood, genes and purity. Lowenthal (1996) encapsulates these themes as 'being first' and 'being innate'.

Emphasis on stability can also perpetuate a very static view of tradition. Working in a Thai context, Byrne cautioned that 'the struggle to achieve management control over the material past is one which takes place not across East–West, national, or cultural lines but across or at the borders of different and often competing discursive formations' (1995: 267). Thus the problem may not be so much Western imperialism as the hegemonic tendencies of archaeological discourse, which uses the language of 'universal' heritage to privilege its own meanings over those of local voices, in this example Thai Buddhism. Byrne outlines the tensions in management of the bell-shaped religious monuments (*stupa*) between attempts to preserve their 'original' fabric – often consistent with the nation-building desires of the state – and 'the ethic of merit making in Thai Buddhism, [which] puts a premium on the proliferation, reconstruction and rebuilding of *stupas*' (p. 267).

9.2.2 Monuments

At the very least, stones which are marked or emplaced in the landscape tell us that someone else was here first. The centrality of monuments in many visions of the past, and their importance within heritage, reflect a variety of social processes (Illustration 9.6). Obviously, stones and rocks physically preserve better than most other aspects of heritage, and thus bias our view of the past to some extent. However, many would argue that the connection between long-lived substances and heritage is no coincidence. Taçon argues that the widespread use of rock marking by prehistoric peoples demonstrates that they recognized the permanency of stone, and thus further shows their sense of past (over which time the stone has survived) and future (when others will observe the marking):

> the overriding implication of any group of people painting, engraving or assembling blocks or

ILLUSTRATION 9.6 Buddha relief on stone, Namsan Mountains Park, near Kyongju, South Korea (by permission of Gordon Waitt)

boulders of stone ... is that stone is associated with a sense of permanency for all peoples. It outlasts individuals and generations in a way that wood, bone or more fragile substances from which artefacts are made never can (but this is not to say that people did not use these and other substances for symbolic purposes alongside stone). The stone walls, ceilings, shelters, caves, platforms and blocks are also significant parts of the landscape, counterparts to the more fragile earth, plant and animal life. (Taçon, 1994)

This is seen in the use of cupule rock art – pecked circular depressions – in northern Australia. Cupules are distinguished from utilitarian grinding hollows on a range of criteria (Taçon *et al.*, 1997), and make visual reference to solutional weathering features in the sandstones (Illustration 9.7). They occur earliest in the regional rock art sequence, and there is little recorded contemporary Aboriginal knowledge

of them. Thus while the meanings attributed to them by their makers will remain obscure, certain trends can be discerned by rigorous study of their placement in the landscape. In a study of 26 cupule localities within the area of their greatest concentration, Taçon *et al.* (1997) recorded at least 10 different ways in which cupules were used to mark places: for example, to define boundaries (e.g. between inside and outside of shelters); link localities via pathways of marked boulders; highlight natural features such as holes, tunnels and passageways; and define spaces within rock shelters. They argue that cupules 'mark and define a human universe at and between particular localities, to accentuate or separate culturally defined areas from the natural' (p. 961).

Much heritage legislation still gives priority to a more narrowly conceived notion of 'monuments', leading to several biases in our understanding of the past (Mulk and Bayliss-Smith, 1999). These include in particular a focus on sites rather than total landscapes.

9.3 Selling culture and the past: landscapes of tourism

The past is ... being re-invented, sanitized, simplified and packaged for tourist consumption by both internal and foreign tourists, and I suspect that this will have more effect in the next few years on the way the past is investigated and presented than any shifts in academic paradigms. (Glover, 1999: 599)

When issues of heritage and identity are entwined with those of international tourism, things become even more complicated. Indeed, Pai (1999: 624) argues that 'the present goal of heritage management in South Korea is to define core-areas for tourism' (*see* Box 9.1). The considerable infrastructure associated with linking cultural heritage sites with commerce is itself a source of environmental change and ecological pressure. While nation-states and heritage organizations are still active agents in representing the past for tourists, they are jostled by a myriad other participants in the marketplace. For environmental managers, tourism is becoming a major issue as affluent travellers look for new undiscovered destina-

ILLUSTRATION 9.7 Cupules of varying ages at the Jinmium rockshelter site, Northern Territory, Australia, extend below the present ground surface. Archaeological excavation aims to relate cupule art to other changes in the prehistoric record (by permission of Richard Fullagar)

tions. Visitor pressures bring the attendant difficulties of 'loving to death' the attributes which attracted them in the first place. By the same token, tourism can be a resource for conservation, in that it encourages investment in the protection of valued sites and landscapes.

9.3.1 Tourism, wilderness and landscapes

As was discussed in earlier chapters, many of the landscapes valued by tourists for their emptiness of human associations are in fact someone's home. There are many examples of the simultaneous construction of these two kinds of cultural

---

Box 9.1 Culture 'core areas' in South Korea, as defined by the Tourist Development Office

The key strategic initiative for the future development of tourism in South Korea is now focused on the promotion of six core areas:

- Seoul (Yi dynastic culture);
- Puyo/Kongju (Paekche culture: first to third centuries);
- Kyongju (Silla culture: third to tenth centuries);
- Chungwon;
- Pusan (Kaya culture: first to third centuries);
- Cheju island.

The selection of the six proposed centres of culture (*munwha-kwon*) was based on their unique cultural properties to the area. The properties were ranked in descending order of importance as:

- national monuments (art-historical sites, architecture, sculpture, archaeological remains and museums);
- lifestyles and subsistence strategies (historical battle sites of foreign resistance such as Chinju-song, the site of anti-Hideyoshi resistance during the Hideyoshi Invasions (1592–97), and the villa of Korea's most famous naval hero, Yi Sun-shin);
- technological and scientific achievements of the Korean people (e.g. turtle-boats at Hansan-do, an island where Yi Sun-shin sank the invading Japanese armada during the Hideyoshi invasions);
- holy sites/sacred places (e.g. Kanghwa-do's shrine of the founding ancestor, Tan'gun);
- sites of myths/legends.

Linked to these projects are long-range infrastructure development plans including new roads, frequent bus and train services, and hotels.

Source: Pai (1999: 624–5)

landscape. To follow the Sámi example from earlier in the chapter, this is seen in the marketing of the Jokkmokk area as 'wilderness country': 'Undisturbed nature is a human right. ... The municipality of Jokkmokk accommodates large tracts of national parks and reserves ... which form a huge wilderness area. Large expanses are needed to give an adequate shelter to wildlife' (quoted in Mulk and Bayliss-Smith, 1999: 366). In the same breath, the Swedish Environmental Protection Agency maintenance plans for Sarek National Park (Figure 9.3) specify both that 'this unique untouched environment is to be preserved intact' (p. 366) and that Sámi reindeer husbandry is still practised, with some Sámi huts surviving.

The desire of adventure-seekers, back-packers and bushwalkers for 'uncultural' landscapes conflicts in this example with Sámi livelihood and symbolic meaning. Virtually identical issues are played out in many other areas, for example Australia and Canada. As we have seen at various points throughout the book, the ground is shifting towards increased recognition of indigenous rights and presence.

It remains to be seen whether Western conservationists can value such landscapes as highly as the natural ones of their imagination. Within a globalized environmental movement and tourism market, the process of 'othering' goes well beyond national boundaries. 'Remote' corners like northern Sweden and central Australia are sites onto which fantasies from many metropolitan centres are projected.

This notion of Australia as Other is actively exploited by the Australian Tourist Commission, which has to define and package an Australian identity and position it in the international tourism market (Waitt, 1997). This process requires Australia to be differentiated from competing products in markets including North America, Japan, Europe and other parts of Asia. 'To meet the demands of adventurers the landscape of the interior (The Red Centre) is symbolised in the visual text [television advertisements] as a wild, rugged, vibrantly red 'wilderness' ... all of course devoid of people (a founding concept of colonial appropriation)' (Waitt, 1997: 50). Representations of indigenous people in the advertisements are as part of the

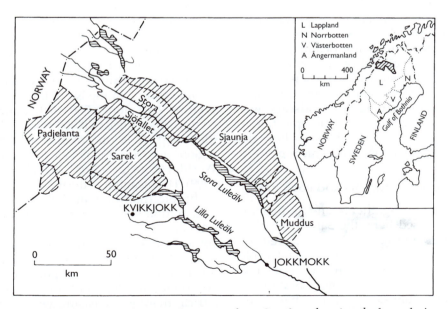

Figure 9.3 Lule River, upper valley, in northern Sweden, showing the boundaries of various national parks which constitute Laponia World Heritage Area
Source: Mulk and Bayliss-Smith (1999: figure 24.1). By permission of Taylor & Francis Books Ltd

landscape, historical and romantic. The visual images of Aborigines used are male, loin-clothed and holding spears.

Where tourist enterprises are owned and controlled by Aboriginal people there is greater scope for contesting 'the tea towel depiction of Aboriginal culture' (Langton *et al.*, 1999: 101). The greatest growth in this area is in specialist niches where people are willing and able to pay for high-quality experiences with Aboriginal people. While such operations still have to operate in a context in which tourist expectations have been created by wider forces, they are at least in a position to contest the 'empty landscape' ideas. Neither is large-scale tourism something that indigenous communities necessarily want to be involved in, many finding it more intrusive than economic alternatives such as resource extraction.

Uluru–Kata Tjuta National Park provides one example of the meeting of mass tourism and indigenous aspirations, with the Board and Draft Plan of Management providing mechanisms for protection all round. In this context of indigenous control the message is a positive one – *Pukulpa Pitjama Ananguku Ngurakutu* – Welcome to Aboriginal Land – notwithstanding the high visitor numbers (339,650 fee-paying visitors in 1998) (Australia, 1999: 95). The strategic approach to visitor planning includes consideration of cultural protection as well as more nitty-gritty issues such as transportation, access and accommodation. In Chapter 7 different ways of seeing this landscape were exemplified in relation to the question of 'climbing the rock'; Box 9.2 shows how that issue is addressed in the management plan.

9.3.2 Tradition, tourism and rock art

Like other hunter-gatherer and indigenous peoples discussed here, the San of the Natal Drakensberg, South Africa, have moved into and out of history's story as it suited later colonizers. In the past century described as everything from 'brutal savages' to 'harmless people' and 'noble savages', they have more recently been promoted as living in harmony with their environment: 'It is ironic that, a century after the San fought and lost their struggle for the Natal Drakensberg, they are being used by descendants of their white settler adversaries to promote the conservation of the area' (Mazel, 1992: 765).

This view of 'traditional' Africans as 'living in the pristine state of an idealized "natural" past' is widespread in tourist literature and postcards in South Africa (Spiegel, 1994: 192), meeting the considerable tourist demand for an experience of Africa as 'other'. Some images of this traditionality are highly commodified. For example, the exquisite detail in the rock paintings translates well into TrueType fonts, which can be bought as a set for Windows 95 or 98 applications. The advertisement tells us, 'some of these images are nearly 28 000 years old ... amongst the oldest known artworks of humankind'. (This widely quoted date is derived from a painted stone excavated from 27,000-year-old archaeological deposits in Namibia. Most surviving paintings are much younger than this (Wahl, 1999).) The global economy and stylish contemporary designs merge seamlessly with the past and notions of universal heritage.

In such a context, management and interpretation of San rock art is a fraught business. Protection of rock art from processes of physical weathering (see Chapter 8) starts to look simple! Hunter-gatherers lived in the *Khahlamba* (Drakensberg) area from about 8000 years ago until British colonization in the nineteenth-century. In the Natal Drakensberg Park (Figure 9.4) alone, 550 rock painting sites comprising at least 40,000 images have been identified. It is one of the richest rock art regions in Africa, and is soon to be proposed for World Heritage status (Mazel, 1992; Wahl, 1999). The meaning of the images, such as theri-anthropic (half-human, half-animal) figures (Illustration 9.8a), is debated among rock art scholars; hunter-gatherer ancestor figures, spirits of the dead, and transformed shamans have all been proposed. As international tourism becomes an important source of economic development in the new South Africa, there is increasing visitor interest in, and pressure on, the rock art sites.

The relatively accessible Main Caves at Giant's Castle Game Reserve currently receive about

Box 9.2   Climbing Uluṟu: cultural and management dilemmas

Although climbing Uluṟu remains a popular activity for some visitors, it is the view of Nguraritja Winki that visitors should not climb. They consider that to climb is to show disrespect for the spiritual and safety aspects of Tjukurpa. They are very concerned about visitor safety. Each time a visitor is seriously or fatally injured at Uluṟu, Nguraritja Winki share in the grieving process. Tjukurpa requires that Nguraritja Winki take responsibility for looking after visitors to their country: this 'duty of care' is the basis of their stress and grieving for those injured. Parks Australia shares these views.

The overall percentage of Park visitors climbing Uluṟu has declined in recent years as visitor awareness of Nguraritja Winki views has increased.

Aims

- to discourage people from climbing Uluṟu;

- to reduce the number of Uluṟu-related safety incidents;

- review and integrate strategies and policies associated with activities in and around Uluṟu.

Actions

- *Climb route*. Climbing will be permitted only on the existing Uluṟu climb and climbers will be required to keep to the marked route.

- *Closures*. As in the past, the climb will be closed if conditions compromise visitors' safety.

- *Review*. Conduct a review of the strategy and policies associated with activities around Uluṟu.

- *Site planning*. Site planning for the base of the climb – Mala Walk track head will aim to change the emphasis away from the climb route. This will involve road and car park relocation, better shade and seating, and more safety information.

- Realignment of the Uluṟu ring road will be subject to detailed planning with a view to moving the road further away from Uluṟu.

- *Alternatives*. Suitable alternative activities will be developed and promoted to visitors.

- *Interpretation*. The message *Nganana Tatintja Wiya* – 'We Never Climb' – will be presented in interpretive brochures, during guided tours and on signs, to encourage people not to climb Uluṟu.

- The tourism industry will be encouraged to provide pre-visit information nationally and internationally, and tour guides will be required to tell clients that climbing Uluṟu is culturally inappropriate.

- *Pre-climb briefings*. A requirement may be introduced during the life of the Plan for all potential climbers to undertake a briefing on cultural and safety matters.

- *Cultural Centre*. Tour operators and independent visitors will be encouraged to visit the Cultural Centre.

- *Monitoring*. Three years into the term of this Plan there will be a review.

Source: Summarized from Draft Plan of Management (Australia, 1999: section 5.2.2, pp. 109–10)

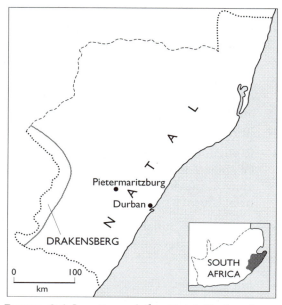

FIGURE 9.4 Location of the Natal Drakensberg, South Africa

18,000 visitors each year. Recently completed large wooden decks, walkways and viewing platforms protect the art by minimizing dust. They also provide visitors with viewing and photographic opportunities (Illustration 9.8b). These sites are also important in that rock art and archaeological deposits occur at the same locations, offering the potential for research to interpret the art in the broader context of prehistoric life. A diorama in South Cave (Illustration 9.8c) has been particularly contentious. Created during the late 1960s, it links the San hunter-gatherers to the prehistoric past by insetting a model of the archaeological deposit beneath their feet. Associated interpretive material emphasizes the interactions of hunter-gatherers with agriculturalists from the surrounding region, and with later European colonists,

> in an effort to dispel the myth that hunter-gatherers lived in isolation. ... When the interpretive centre was upgraded recently we decided to retain the diorama and use it to show interaction and the tragic demise of the hunter-gatherers in the *Khahlamba*. (Wahl, 1999: 7)

Good heritage interpretation can place itself, as well as the site or feature, in historical context.

On the one hand, the concept of a diorama might seem dated and detract from a 'natural' experience of the site (or as natural as you can have from a shared boardwalk). However, by drawing attention to itself and the context in which it was created, it performs two important tasks. It illustrates that interpretations of the past and of hunter-gatherer peoples have changed in the past few decades, and thus encourages visitors to remember that our contemporary interpretations will be revised in the future. Second, by raising the question of *how* interpretations of the past are made, it takes us back to the archaeological evidence itself as a basis for those changing understandings.

9.4 Educating the educators

Most conservation managers have training in the environmental and natural sciences rather than the human sciences. Trained in a tradition of 'truth' and 'facts', they may feel, and be, ill-equipped to deal with landscapes of multiple meanings. Even for heritage professionals trained in the humanities, it is a considerable challenge to think clearly about a physical landscape which is simultaneously

- a product of the activity of different prehistoric peoples over thousands of years, all investing the land and its past with their own meanings;
- valued in the present by diverse peoples for different reasons, e.g. indigenous, scientific, conservationist, tourist; and
- subject to physical pressures that are increasing precisely because it is highly valued.

A variety of training and management strategies are needed. Exposure to the research process at all levels of scientific education will help students understand the contingent nature of science itself, and give them a more dynamic understanding of how scientific knowledge is produced. Managers who have themselves undertaken postgraduate research will be particularly valuable in this regard. Increasingly, environmental scientists receive interdisciplinary

training. This is important for making them able to consider cross-cultural issues and changing historical contexts when making environmental decisions. A decision-maker who understands long-term changes in the landscape will not waste energy trying to reconstruct a single 'authentic' past, but will be aware that the process of valuing cultural, or natural, heritage must be an explicit one.

There are also relevant practical skills. Systematically observing visitor behaviour at rock art sites and negotiating with traditional owners will be just as important as – in the first instance more important than – measuring dust or humidity levels (Box 9.3).

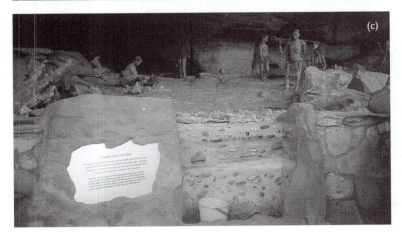

ILLUSTRATION 9.8
(a) Therianthropes, North Cave, Main Caves, Giant's Castle Game Reserve, South Africa
(b) Walkway and viewing platforms, North Cave, Main Caves, Giant's Castle Game Reserve, South Africa
(c) Interpretive diorama, South Cave, Main Caves, Giant's Castle Game Reserve, South Africa

Box 9.3   Visitor management strategies for cultural heritage sites

This overview of strategies is not presented as either a comprehensive overview or a management blueprint. Each cultural heritage site and area needs to be managed in its own context. These strategies are most relevant to Aboriginal heritage sites in Australia, but the issues raised have broader application.

Indirect management strategies

- Protection by heritage legislation and planning provision
- Protection by land tenure
- Private
- Public
- Aboriginal ownership
- Visible and invisible sites – active promotion of some sites; others made 'invisible' by removing from tourist maps, blocking roads, revegetation.

Direct management strategies

Direct management strategies come into play once a site has been selected as public or open. How to minimize and control visitor impacts?

Physical mechanisms
- car parks – ensure only pedestrian traffic reaches site, reduce dust;
- pathways – space out visitors, direct attention to broader context of site, access for elderly and disabled;
- boardwalks and paving – reduce dust and erosion, keep visitors away from rock art faces, photographic opportunities important;

- grouping areas – seating and interpretive material, bays for guides to deliver talks;
- barrier structures – cages, grilles, low wooden fencing, boulder lines;
- signposts – polite requests or strong warnings? Deliberate touching shown to be reduced by introduction of signs.

Financial mechanisms
- ticketing or pricing – distinguish between revenue-raising and pricing for management purposes. Enables an official to be at site;
- sales outlets – allow people to take something home without damaging site;
- photography – confidentiality and cultural property issues. Studies show that people with cameras generally behave better, provided their needs are catered for.

Educative mechanisms
- publicity and promotion – quality is important;
- name changes – renaming as a direct expression of Aboriginal interests, e.g. renaming of Ayers Rock as Uluru;
- visitor centres – very important source of information, but need to enhance the experience of a site, not replace it;
- on-site interpretive signs – must be simple, clear, precise, concise;
- maps and brochures – offer self-regulatory mechanisms in unstaffed areas;
- guides – employment avenue for indigenous owners; not always feasible in rural or remote areas.

Source: After Jacobs and Gale (1994)

10

Conclusion

Like Michael Leunig's characters (Figure 10.1), many of us don't quite know what to make of the past or future, but sense that it is important at least to look in their respective directions. We are more comfortable in the present, which in the end is all we have anyway, finding physical and psychological sustenance in our connec- tions to something we call the natural world. Leunig refashions past and future into 'pasture' – a physical place, but also something else again. Our present world carries both an inher- itance from the past, and the seeds of the future. Similarly, but somewhat less playfully, contemporary environmental management

FIGURE 10.1 'Past and future', by Michael Leunig
As published in the *Sydney Morning Herald*, March 2000. By permission of Michael Leunig/*The Age*

dilemmas require us to deal with a nature that is being refashioned both physically and conceptually. The first challenge for our thinking is to accommodate both change and stability over a range of temporal scales. As examples through the book have shown, this is an important issue whatever the field of inquiry, and whether or not humans are included in the frame of reference.

As we have seen, a number of concepts in environmental management, for example *heritage*, carry within them apparently contra-dictory ideas. *Tradition* is another, containing a 'paradoxical marriage between continuity and change' (Mearns, 1994: 266). The anthropol-ogist Mearns discusses the dilemmas faced by Aboriginal women in Australia's Northern Territory as they seek to fulfil their responsibil-ities towards sacred sites in a changing socio-economic world and in interaction with Australian federal and Territory legislation. The central of these dilemmas is the need to entrust secret knowledge of sites to an outside body (the Aboriginal Areas Protection Authority) – thus potentially breaching Aboriginal law – in order to protect the sites. The process of interaction involves a reorien-tation of the gender relationships under which much ceremonial business is structured. Knowledge and relationships are also trans-formed by the process of making permanent written records of business that has previously been transmitted between the generations by oral and ceremonial means. 'To continue tradition thus means to change it' (Mearns, 1994: 286). Similarly, the Yolngu of Arnhem Land use the expression 'same but different' to discuss variation, particularly of artistic styles, within a stable cosmology (Caruana and Lendon, 1997: 37).

The metaphor of tradition as 'an active and continuous selection and reselection, which even at its latest point in time is always a set of specific choices' (Williams, 1980: 16), may be helpful in relation to our academic traditions of thinking about human–environment interac-tions. What is worth preserving in past ways of thinking, and what needs renovating? For

Raymond Williams, literary 'tradition' provided a common example of a 'false totality' into which the 'facts of change' are projected, contained and made

> at last, like the rocks, [to] stand still. Except, of course, that in the actual physical sciences we soon learn, even against everyday experience, that only some rocks stand still, and that even these are the products of change: the continuing history of the earth. It is not really from science, but from certain philosophical and ideological systems, and I suppose ultimately from religions, that these apparent totalities, which contain, override or rationalize change, are projected. (Williams, 1980: 15–16)

The nature of change in the physical Earth is much better understood now than it was in 1971, when Williams first wrote these words. However, the notion of tradition as being suspended apart from time – as opposed to an active and continuous set of choices firmly embedded within history – has been more persistent.

I have identified throughout the book some points of convergence between the sciences and humanities in thinking about change and continuity. Some of our most traditional empirical tools, such as appropriate scale, are vindicated. Others, such as environmental agency, require some rehabilitation to work off the lingering effects of environmental deter-minism. This is an urgent task because the constructivist critique – important in so many ways – flounders somewhat when it meets the biophysical landscape. Today's students will be well placed to take on these challenges if they have been trained in transdisciplinary thinking and have multidisciplinary skills.

They will perhaps also be the generation of thinkers who can take on the more profound and difficult task with which I began: to rethink people into nature in such a way that we can better manage the Earth. Intellectual weapons which conceptualize humans and nature sepa-rately are in tension with the evidence that human influences are now embedded in virtually all Earth processes. The challenge for all of us is to make this tension a creative one.

REFERENCES

Adams, W.M. 1997: Rationalization and conservation: ecology and the management of nature in the United Kingdom. *Transactions of the Institute of British Geographers* **22**: 277–91.

Agnew, J.A. and **Duncan, J.S.** (eds) 1989: *The power of place*. Boston, Unwin Hyman.

Allen, H. 1997: Conceptions of time in the interpretation of the Kakadu landscape. In Rose, D.B. and Clarke, A. (eds), *Tracking knowledge in north Australian landscapes*. Darwin, North Australia Research Unit, 141–54.

Anderson, A. and **McGlone, M.S.** 1992: Living on the edge: prehistoric land and people in New Zealand. In Dodson, J.R. (ed.), *The naive lands: prehistory and environmental change in Australia and the southwest Pacific*. Melbourne, Longman Cheshire: 199–241.

Anderson, E. 1952: *Plants, man, and life*. Berkeley, University of California Press.

Anderson, K. 1997: A walk on the wild side: a critical geography of domestication. *Progress in Human Geography* **21**: 463–85.

Anderson, K. and **Gale, F.** 1992: Introduction. In Anderson, K. and Gale, F. (eds), *Inventing Places: studies in cultural geography*. Melbourne, Longman Cheshire: 1–14.

Archer, M., Hand, S. and **Godthelp, H.** 1994: *Riversleigh: the story of animals in ancient rainforests of inland Australia*. Sydney, Reed.

Aronson, J., Dhillion, S. and **Le Floc'h, E.** 1995: On the need to select an ecosystem of reference, however imperfect: a reply to Pickett and Parker. *Restoration Ecology* **3**: 1–3.

Aronson, J. and **Le Floc'h, E.** 1996: Hierarchies and landscape history: dialoguing with Hobbs and Norton. *Restoration Ecology* **4**: 327–333.

Australia, Commonwealth of 1994: *Renomination of Uluṟu–Kata Tjuṯa National Park by the Government of Australia for Inscription on the World Heritage List*. Canberra, Department of the Environment, Sport and Territories.

Australia, Commonwealth of 1999: *Uluṟu–Kata Tjuṯa National Park Draft Plan of Management*. Canberra, Uluṟu–Kata Tjuṯa Board of Management/Parks Australia.

Baker, L., Woenne-Green, S. and **Muṯitjulu community** 1993a: Aṉangu knowledge of vertebrates and the environment. In Reid, J.R.W., Kerle, J.A. and Morton, S.R. (eds), *Uluṟu fauna: the distribution and abundance of vertebrate fauna of Uluṟu (Ayers Rock–Mount Olga) National Park, N.T.* Canberra, Australian National Parks and Wildlife Service: 79–120.

Baker, L.M. and **Muṯitjulu community** 1992: Comparing two views of the landscape: Aboriginal traditional ecological knowledge and modern scientific knowledge. *Rangeland Journal* **14**: 174–89.

Baker, R.G., Schwert, D.P., Bettis, E.A. and **Chumbley, C.A.** 1993b: Impact of Euro-American settlement on a riparian landscape in northeast Iowa, midwestern USA: an integrated approach based on historical evidence, flood-plain sediments, fossil pollen, plant macrofossils and insects. *The Holocene* **3**: 314–23.

Ballard, C. 1994: The centre cannot hold: trade networks and sacred geography in the Papua New Guinea highlands. *Archaeology in Oceania* **29**: 130–48.

Barbour, M.G. 1996: Ecological fragmentation in the fifties. In Cronon, W. (ed.), *Uncommon ground: rethinking the human place in nature*. New York, W.W. Norton & Co.: 233–55.

Barham, A.J. 1999: The local environmental impact of prehistoric populations on Saibai island, northern Torres Strait, Australia: enigmatic evidence from Holocene swamp lithostratigrahic records. *Quaternary International* **59**: 71–106.

Barrett, J.C. 1994: *Fragments from antiquity: an archaeology of social life in Britain, 2900–1200 BC*. Oxford, Blackwell.

Barrett, J.C. 1999: Chronologies of landscape. In Ucko, P.J. and Layton, R. (eds), *The archaeology and anthropology of landscape: shaping your landscape*. London, Routledge: 21–30.

Battarbee, R.W. 1990: The causes of lake acidification, with special reference to the role of acid deposition. *Philosophical Transactions of the Royal Society of London B* **327**: 339–47.

Battarbee, R.W. 1991: Recent paleolimnology and diatom-based environmental reconstruction. In Shane, C.K. and Cushing, E.J. (eds), *Quaternary Landscapes*. Minneapolis, University of Minnesota Press: 129–74.

Battarbee, R.W., Flower, R.J., Stevenson, A.C. and **Rippey, B.** 1985: Lake acidification in Galloway: a palaeoecological test of competing hypotheses. *Nature* **314**: 350–2.

Bayliss-Smith, T. and **Golson, J.** 1999: The meaning of ditches: deconstructing the social landscapes of New Guinea, Kuk, Phase 4. In Gosden, C. and Hather, J. (eds), *The prehistory of food: appetites for change*. London, Routledge: 199–231.

Behre, K.-E. 1981: The interpretation of anthropogenic indicators in pollen diagrams. *Pollen et Spores* **23**: 225–45.

Behre, K.-E. 1988: The role of man in European vegetation history. In Huntley, B and Webb, T. (eds) *Vegetation History*. Dordrecht, Kluwer: 633–72.

Bell, M. and **Evans, D.M.** 1997: Greening 'the heart of England': redemptive science, citizenship, and 'symbol of hope for the nation'. *Environment and Planning D: Society and Space* **15**: 257–79.

Bell, M. and **Walker, M.J.C.** 1992: *Late Quaternary environmental change: physical and human perspectives*. Harlow, Longman.

Bender, B. (ed.) 1993a: *Landscape: politics and perspectives*. Oxford, Berg.

Bender, B. 1993b: Stonehenge: contested landscapes (medieval to present-day). In Bender, B. (ed.), *Landscape: politics and perspectives*. Oxford, Berg: 245–80.

Bender, B. 1999: Introductory comments on 'Dynamic landscapes and socio-political process: the topography of anthropogenic environments in global perspective'. *Antiquity* **73**: 632–4.

Berg, L.D. and **Kearns, R.A.** 1996: Naming as norming: 'race', gender, and the identity politics of naming places in Aotearoa/New Zealand. *Environment and Planning D: Society and Space* **14**: 99–122.

Birks, H.H., Birks, H.J.B., Kaland, P.E. and **Moe, D.E.** (eds) 1988: *The cultural landscape: past, present and future*. Cambridge, Cambridge University Press.

Bishop, P. (ed.) 1988: *Lessons for human survival: nature's record from the Quaternary*. Sydney, Geological Society of Australia Symposium Proceedings 1.

Blaikie, P.M. 1996: Post-modernism and global environmental change. *Global Environmental Change* **6**: 81–85.

Blumler, M.A. 1998: Biogeography of land-use impacts in the Near East. In Zimmerer, K.S. and Young, K.R. (eds), *Nature's geography: new lessons for conservation in developing countries*. Madison, University of Wisconsin Press: 215–36.

Blunden, J. and **N. Curry** (eds) 1990: *A people's charter? Forty years of the National Parks and Access to the Countryside Act 1949*. London, Her Majesty's Stationery Office.

Bocking, S. 1997: *Ecologists and environmental politics: a history of contemporary ecology*. New Haven, Yale University Press.

Bodmer, R.E., Eisenberg, J.F. and **Redford, K.H.** 1997: Hunting and the likelihood of extinction of Amazonian mammals. *Conservation Biology* **11**: 460–6.

Bomford, M. and **Caughley, J.** (eds) 1996: *Sustainable use of wildlife by Aboriginal peoples and Torres Strait Islanders*. Canberra, Australian Government Publishing Service.

Bond, G.C. and **Gilliam, A.** (eds) 1994: *Social construction of the past: representation as power*. London, Routledge.

Bonnieux, F. and **Le Goffe, P.** 1997: Valuing the benefits of landscape restoration: a case study of the Cotentin in Lower-Normandy, France. *Journal of Environmental Management* **50**: 321–33.

Botkin, D.B. 1990: *Discordant harmonies: a new ecology for the twenty-first century*. Oxford, Oxford University Press.

Bottema, S., Entjes-Nieborg, G. and **Van Zeist, W.** (eds) 1990: *Man's role in the shaping of the Eastern Mediterranean landscape*. Rotterdam, Balkema.

Bradley, R. 1984: *The social foundations of prehistoric Britain*. London, Longman.

Bradley, R. 1997: Domestication as a state of mind. *Analecta Praehistorica Leidensia* **29**: 13–18.

Brantenberg, O.T. 1991: Norway: constructing indigenous self-government in a nation state. In Jull, P. and Roberts, S.R. (eds), *The challenge of northern regions*. Darwin, North Australia Research Unit: 66–128.

Brookes, A. 1995: River channel restoration: theory and practice. In Gurnell, A. and Petts, G. (eds) *Changing River Channels*. Chichester, John Wiley: 369–88.

Brosius, J.P. 1986: River, forest and mountain: the Penan Gang landscape. *Sarawak Museum Journal* **36** (57, New Series): 173–84.

Brosius, J.P. 1997: Endangered forest, endangered people: environmentalist representations of indigenous knowledge. *Human Ecology* **25**: 47–68.

Brower, B. and **Dennis, A.** 1998: Grazing the forest, shaping the landscape? Continuing the debate about forest dynamics in Sagarmatha National Park. In Zimmerer, K.S. and Young, K.R. (eds), *Nature's geography: new lessons for conservation in developing countries*. Madison, University of Wisconsin Press: 184–208.

Brown, A.G. 1996: Human dimensions of palaeohydrological change. In Branson, J., Brown, A.G. and Gregory, K.J. (eds), *Global Continental Changes: the Context of Palaeohydrology*. Geological Society Special Publication **115**: 57–72.

Brown, A.G. 1997a: Clearances and clearings: deforestation in Mesolithic/Neolithic Britain. *Oxford Journal of Archaeology* **16**(2): 133–46.

Brown, A.G. 1997b: *Alluvial geoarchaeology: floodplain archaeology and environmental change*. Cambridge, Cambridge University Press.

Brown, A.G. 1999: Characterising prehistoric lowland environments using local pollen assemblages. In Edwards, K. and Sadler, K. (eds), *Holocene environments of prehistoric Britain*. Quaternary Proceedings 7. Chichester, John Wiley.

Brown, A.G. and **Quine, T.A.** (eds) 1999: *Fluvial processes and environmental change*. Chichester, John Wiley.

Burchard, I. 1998: Anthropogenic impact on the climate since man began to hunt. *Palaeogeography, Palaeoclimatology, Palaeoecology* **139**: 1–14.

Burney, D.A., James, H.F., Grady, F.V., *et al.* 1997: Environmental change, extinction and human activity: evidence from caves in NW Madagascar. *Journal of Biogeography* **24**: 755–67.

Byrne, D. 1995: Buddhist *stupa* and Thai social practice. *World Archaeology* **27**: 266–81.

Cairns, J. 1991: The status of the theoretical and applied science of restoration ecology. *The Environmental Professional* **13**: 186–94.

Caruana, W. and **Lendon, N.** (eds) 1997: *The painters of the Wagilag Sisters story 1937–1997*. Canberra, National Gallery of Australia.

Caseldine, C. and **Hatton, J.** 1994: The development of high moorland on Dartmoor: fire and the influence of Mesolithic activity on vegetation change. In Chambers, F.M. (ed.), *Climate change and human impact on the landscape*. London, Chapman & Hall: 119–31.

Chambers, F. M. (ed.) 1994: *Climate change and human impact on the landscape*. London, Chapman & Hall.

Chase, A. 1989: Domestication and domiculture in northern Australia: a social perspective. In Harris, D.R. and Hillman, G.C. (eds), *Foraging and farming: the evolution of plant exploitation*. London, Unwin Hyman: 42–78.

Clark, R.L. 1983: Pollen and charcoal evidence for the effects of Aboriginal burning on the vegetation of Australia. *Archaeology in Oceania* **18**: 32–7.

Cloke, P., Milbourne, P. and **Thomas, C.** 1996: The English National Forest: local reactions to plans for renegotiated nature–society relations in the countryside.

Transactions of the Institute of British Geographers **21**: 552–71.

Colinvaux, P.A., Bush, M.B., Steinitz-Kannan, M. and **Miller, M.C.** 1997: Glacial and postglacial pollen records from Ecuadorian Andes and Amazon. *Quaternary Research* **48**: 69–78.

Colinvaux, P.A., De Oliveira, P.E., Moreno, J.E., Miller, M.C. and **Bush, M.B.** 1996: A long pollen record from lowland Amazonia: forest and cooling in glacial times. *Science* **274**: 85–8.

Colley, S. 1995: What happened at WAC-3? *Antiquity* **69**: 15–18.

Connell, J.H. 1978: Diversity in tropical rainforests and coral reefs. *Science* **199**: 1302–10.

Cooke, R. 1998: Human settlement of central America and northernmost South America (14,000–8000 BP). *Quaternary International* **49/50**: 177–90.

Cooney, G. 1999: Social landscapes in Irish prehistory. In Ucko, P.J. and Layton, R. (eds), *The archaeology and anthropology of landscape: shaping your landscape.* London, Routledge: 46–64.

Cooney, G. and **Grogan, E.** 1998: People and place during the Irish Neolithic: exploring social change in time and space. In Edmonds, M. and Richards, C. (eds), *Understanding the Neolithic of North-western Europe.* Glasgow, Cruithne Press: 456–80.

Cooper, R. 1995: *Maori customary use of native birds, plants and other traditional materials.* Report presented to Annual Conference of the Australasian Wildlife Management Society, Christchurch.

Cosgrove, D. 1990: Environmental thought and action: pre-modern and post-modern. *Transactions of the Institute of British Geographers* NS **15**: 344–58.

Cosgrove, D. and **Daniels, S.** (eds) 1988: *The iconography of landscape: essays on the symbolic representation, design and use of past environments.* Cambridge, Cambridge University Press.

Cronon, W. 1983: *Changes in the land: Indians, colonists, and the ecology of New England.* New York, Hill & Wang.

Cronon, W. 1991: *Nature's metropolis: Chicago and the Great West.* New York, W.W. Norton.

Cronon, W. 1996a: The trouble with wilderness; or, getting back to the wrong nature. In Cronon, W. (ed.), *Uncommon ground: rethinking the human place in nature.* New York, W.W. Norton: 69–90.

Cronon, W. (ed.) 1996b: *Uncommon ground: rethinking the human place in nature.* New York, Norton.

Dahm, C.N., Cummins, K.W., Valett, H.M. and **Coleman, R.L.** 1995: An ecosystem view of the restoration of the Kissimmee River. *Restoration Ecology* **3**: 225–38.

Daniels, S. 1989: Marxism, culture, and the duplicity of landscape. In Peet, R. and Thrift, N. (eds), *New Models in Geography: the political-economy perspective.* London, Unwin Hyman: 196–220.

Davis, M.B. 1983: Holocene vegetational history of the Eastern United States. In Wright, H.E.J. (ed.), *Late-Quaternary environments of the United States*, vol. 2, *The Holocene.* Minneapolis, University of Minnesota Press: 166–81.

Davis, M.B. 1994: Ecology and palaeoecology begin to merge. *Trends in Ecology and Evolution* **9**: 357–8.

Davis, W. 1990: A way to stay. In Davis, W. and Henley, T. (eds), *Penan: voice for the Borneo Rainforest Committee.* Vancouver, Western Canada Wilderness Committee: 97–101.

Dear, M. 1986: Postmodernism and planning. *Environment and Planning D: Society and Space* **4**: 367–84.

DeBusk, G.H. 1998: A 37,500-year pollen record from Lake Malawi and implications for the biogeography of afromontane forests. *Journal of Biogeography* **25**: 479–500.

Decher, J. 1997: Conservation, small mammals, and the future of sacred groves in West Africa. *Biodiversity and Conservation* **6**: 1007–26.

Delcourt, H.R. 1987: The impact of prehistoric agriculture and land occupation on natural vegetation. *Trends in Ecology and Evolution* **2**: 39–44.

Delcourt, P.A. and **Delcourt, H.R.** 1987: *Long-term forest dynamics of the temperate zone: a case study of late-Quaternary forests in eastern North America.* New York, Springer-Verlag.

Demeritt, D. 1994: The nature of metaphors in cultural geography and environmental history. *Progress in Human Geography* **18**: 163–85.

Demeritt, D. 1996: Social theory and the reconstruction of science and geography. *Transactions of the Institute of British Geographers* NS **21**: 484–503.

Demeritt, D. 1998: Science, social constructivism and nature. In Braun, B. and Castree, N. (eds), *Remaking reality: nature at the millennium.* London, Routledge: 173–93.

Denevan, W.M. 1992: The pristine myth: the landscape of the Americas in 1492. *Annals of the Association of American Geographers* **82**(3): 369–85.

Dodson, J.R. (ed.) 1992: *The naive lands: prehistory and environmental change in Australia and the south-west Pacific.* Melbourne, Longman Cheshire.

Dodson, J.R. and **Intoh, M.** 1999: Prehistory and palaeoecology of Yap, federated states of Micronesia. *Quaternary International* **59**: 17–26.

Dominy, M.D. 1995: White settler assertions of native status. *American Ethnologist* **22**: 358–74.

Downs, P.W. and **Thorne, C.R.** 1998: Design principles and suitability testing for rehabilitation in a flood defence channel: the River Idle, Nottinghamshire, UK. *Aquatic Conservation: Marine and Freshwater Ecosystems* **8**: 17–38.

Drury, W.H. 1998: *Chance and change: ecology for conservationists.* Berkeley, University of California Press.

Duffy, R. 1997: The environmental challenge to the nation-state: superparks and national parks policy in Zimbabwe. *Journal of Southern African Studies* **23**: 441–51.

Duncan, J. and **Duncan, N.** 1988: Rereading the landscape. *Environment and Planning D: Society and Space* **6**: 117–26.

Dunning, N., Scarborough, V., Valdez, F., Luzzadder-Beach, S., Beach, T. and **Jones, J.G.** 1999: Temple mountains, sacred lakes, and fertile fields: ancient Maya landscapes in northwestern Belize. *Antiquity* **73**: 650–60.

Edmonds, M. and **Richards, C.** (eds) 1998: *Understanding the Neolithic of North-western Europe.* Glasgow, Cruithne Press.

Ehlers, E. 1999: Environment and Geography: International Programs on Global Environmental Change. *International Geographical Union Bulletin* **49**: 5–18.

Enright, N.J. and **Gosden, C.** 1992: Unstable archipelagos: south-west Pacific environment and prehistory since 30,000 BP. In Dodson, J.R. (ed.), *The naive lands: prehistory and environmental change in Australia and the southwest pacific.* Melbourne, Longman Cheshire: 160–198.

Erickson, C.L. 1999: Neo-environmental determinism and agrarian 'collapse' in Andean prehistory. *Antiquity* **73**: 634–42.

Evans, J.G. 1994: The influence of human communities on the English chalklands from the Mesolithic to the Iron Age: the molluscan evidence. In Chambers, F.M. (ed.), *Climate change and human impact on the landscape.* London, Chapman & Hall: 147–56.

Evernden, N. 1992: *The social creation of nature.* Baltimore, Johns Hopkins University Press.

Faegri, K. 1988: Preface. *The cultural landscape: past, present and future.* In Birks, H.H., Birks, H.J.B., Kaland, P.E. and Moe, D. (eds), Cambridge, Cambridge University Press: 1–6.

Fairhead, J. and **Leach, M.** 1996: *Misreading the African landscape.* Cambridge, Cambridge University Press.

Fairhead, J. and **Leach, M.** 1998: *Reframing deforestation. Global analyses and local realities: studies in West Africa.* London, Routledge.

Feit, H.A. 1994: The enduring pursuit: land, time and social relationships in anthropological models of hunter-gatherers and in subarctic hunters' images. In Burch, E.S. and Ellanna, J. (eds), *Key issues in hunter-gatherer research.* Oxford, Berg: 421–40.

FitzSimmons, M. and **Goodman, D.** 1998: Incorporating nature: environmental narratives and the reproduction of food. In Braun, B. and Castree, N. (eds) *Remaking reality: nature at the millennium.* London, Routledge: 194–220.

Flannery, T.F. 1990: Pleistocene faunal loss: implications of the aftershock for Australia's past and future. *Archaeology in Oceania* 25: 45–76.

Flannery, T.F. 1994: *The future eaters: an ecological history of the Australasian lands and people.* Sydney, Reed.

Flenley, J.R. 1988: Palynological evidence for land use changes in South-East Asia. *Journal of Biogeography* 15: 185–97.

Flenley, J.R. 1994: Pollen in Polynesia: the use of palynology to detect human activity in the Pacific islands. In Hather, J. (ed.), *Tropical archaeobotany: applications and new developments.* London, Routledge: 202–14.

Foster, D. 1992: Applying the Yellowstone model in America's backyard: Alaska. In Birckhead, J., De Lacy, T. and Smith, L. (eds) *Aboriginal involvement in parks and protected areas.* Canberra, Aboriginal Studies Press: 363–76.

Fullagar, R. and **Head, L.** 1999: Exploring the prehistory of hunter-gatherer attachments to place: an example from the Keep River area, Northern Territory, Australia. In Ucko, P.J. and Layton, R. (eds), *The archaeology and anthropology of landscape: shaping your landscape.* London, Routledge: 322–35.

Fullagar, R.L.K., Price, D.M., Head, L.M. 1996: Early human occupation of northern Australia: archaeology and thermoluminescence dating of Jinmium rock-shelter, Northern Territory. *Antiquity* 70: 751–73.

Gartner, W.G. 1999: Late woodland landscapes of Wisconsin: ridged fields, effigy mounds and territoriality. *Antiquity* 73: 671–83.

Ghaffar, A. and **Robinson, G.M.** 1997: Restoring the agricultural landscape: the impact of government policies in East Lothian, Scotland. *Geoforum* 28: 205–17.

Glacken, C.J. 1967: *Traces on the Rhodian shore: nature and culture in Western thought from ancient times to the end of the eighteenth century.* Berkeley, University of California Press.

Glover, I.C. 1999: Letting the past serve the present: some contemporary uses of archaeology in Viet Nam. *Antiquity* 73: 594–601.

Golson, J. 1995: What went wrong with WAC 3 and an attempt to understand why. *Australian Archaeology* 41: 48–54.

Goodwin, C.N., Hawkins, C.P., Kershner, J.L. *et al.* 1997: Riparian restoration in the western United States: overview and perspective. *Restoration Ecology* 5: 4–14.

Gore, J.A. (ed.) 1985: *The restoration of rivers and streams: theories and experience.* Boston, Butterworth.

Gosden, C. 1989: Prehistoric social landscapes of the Arawe Islands, west New Britain province, Papua New Guinea. *Archaeology in Oceania* 24: 45–58.

Gosden, C. 1994: *Social being and time.* Oxford, Blackwell.

Gosden, C. and **Head, L.** 1994: Landscape: a usefully ambiguous concept. *Archaeology in Oceania* 29: 113–16.

Gosden, C. and **Head, L.** 1999: Different histories: a common inheritance for Papua New Guinea and Australia. In Gosden, C. and Hather, J. (eds), *The prehistory of food: appetites for change.* London, Routledge: 232–51.

Gosden, C. and **Lock, G.** 1998: Prehistoric histories. *World Archaeology* 30: 2–12.

Gosden, C. and **Pavlides, C.** 1994: Are islands insular? Landscape vs. seascape in the case of the Arawe Islands, Papua New Guinea. *Archaeology in Oceania* 29: 162–71.

Goudie, A. 1993: *The human impact on the natural environment.* Oxford, Blackwell.

Graetz, R.D. 1988: Kangaroos, sheep and the rangelands: the ecological issues. *Australian Zoologist* 24(3): 137–8.

Graham, B., Ashworth, G.J. and **Tunbridge, J.E.** 2000: *A geography of heritage: power, culture and economy.* London, Arnold.

Haberle, S. 1998a: Dating the evidence for agricultural change in the highlands of New Guinea: the last 2000 years. *Australian Archaeology* 47: 1–19.

Haberle, S.G. 1993: Late Quaternary Environmental history of the Tari Basi, Papua New Guinea. PhD thesis, Department of Biogeography and Geomorphology. Canberra, Australian National University.

Haberle, S.G. 1994: Anthropogenic indicators in pollen diagrams: problems and prospects for late Quaternary palynology in New Guinea. In Hather, J. (ed.), *Tropical archaeobotany: applications and new developments.* London, Routledge: 172–201.

Haberle, S.G. 1998b: Can climate shape cultural development? Resource Management in Asia-Pacific Working Paper 18, Australian National University, Canberra.

Haberle, S.G., Hope, G.S. and **DeFretes, Y.** 1991: Environmental changes in the Baliem Valley, montane Irian Jaya, Republic of Indonesia. *Journal of Biogeography* 18: 25–40.

Haberle, S.G., Hope, G.A. and **Van der Kaars, S.** (in press). Biomass burning in Indonesia and Papua New Guinea: decoupling human and natural fire events in the fossil record. *Palaeogeography, Palaeoclimatology, Palaeoecology.*

Habu, J. and **Fawcett, C.** 1999: Jomon archaeology and the representation of Japanese origins. *Antiquity* 73: 587–93.

Hagen, J.B. 1992: *An entangled bank: the origins of ecosystem ecology.* New Brunswick, NJ, Rutgers University Press.

Haraway, D. 1991: *Simians, cyborgs, and women.* London, Free Association Books.

Harding, S. 1986: *The science question in feminism.* Ithaca, NY, Cornell University Press.

Harvey, D. 1989: *The condition of postmodernity*. Oxford, Blackwell.

Head, L. 1989: Prehistoric aboriginal impacts on Australian vegetation: an assessment of the evidence. *Australian Geographer* **20**: 37–46.

Head, L. 1990: Conservation and Aboriginal land rights: when green is not black. *Australian Natural History* **23**: 448–54.

Head, L. 1994a: Both ends of the candle? Discerning human impact on the vegetation. *Australian Archaeology* **39**: 82–6.

Head, L. 1994b: Landscapes socialised by fire: post-contact changes in Aboriginal fire use in northern Australia, and implications for prehistory. *Archaeology in Oceania* **29**: 172–81.

Head, L. 1996: Rethinking the prehistory of hunter-gatherers, fire and vegetation change in northern Australia. *The Holocene* **6**: 481–7.

Head, L. 2000: *Second Nature: the history and implications of Australia as Aboriginal landscape*. New York, Syracuse University Press.

Head, L., Gosden, C. and **White, J.P.** 1994: Social landscapes. *Archaeology in Oceania* **29**(3).

Head, L. and **Hughes, C.** 1996: One land, which law? Fire in the Northern Territory. In Howitt, R., Connell, J. and Hirsch, P. (eds), *Resources, nations and Indigenous peoples: case studies from Australasia, Melanesia and Southeast Asia*. Melbourne, Oxford University Press: 278–88.

Herricks, E.E. and **Osborne, L.L.** 1985: Water quality restoration and protection in steams and rivers. In Gore, J.A. (ed.), *The restoration of rivers and streams: theories and experience*. Boston, Butterworth: 1–20.

Higgs, E.S. 1997: What is good ecological restoration? *Conservation Biology* **11**: 338–48.

Hirsch, E. and **O'Hanlon, M.** (eds) 1995: *The anthropology of landscape: perspectives on place and space*. Oxford, Clarendon Press.

Hobbs, R.J. and **Norton, D.A.** 1996: Towards a conceptual framework for restoration ecology. *Restoration Ecology* **4**: 93–110.

Hodder, I. 1990: *The domestication of Europe*. Oxford, Blackwell.

Hodell, D.A., Curtis, J.H. and **Brenner, M.** 1995: Possible role of climate in the collapse of Classic Maya civilization. *Nature* **375**: 391–4.

Holdaway, R.N. 1996: Arrival of rats in New Zealand. *Nature* **384**: 225–226.

Hooghiemstra, H. and **Van der Hammen, T.** 1998: Neogene and Quaternary development of the Neotropical rainforest: the forest refugia hypothesis, and a literature overview. *Earth Science Reviews* **44**: 147–83.

Hope, G.S. 1999a: *The anthropogenic impact of human settlement in Fiji*. 15th International INQUA Congress, Durban, South Africa.

Hope, G.S. 1999b: Vegetation and fire response to late Holocene human occupation in island and mainland north west Tasmania. *Quaternary International* **59**: 47–60.

Horrocks, M., Ogden, J., Nichol, S.L., Alloway, B.V., Sutton, D.G. 2000. Palynology, sedimentology and environmental significance of Holocene swamps at northern Kaitoke, Great Barrier Island, New Zealand. *Journal of the Royal Society of New Zealand* **30**: 27–47.

Horton, D. 2000: *The pure state of nature: sacred cows, destructive myths and the environment*. Sydney: Allen & Urwin.

Horton, D.R. 1982: The burning question: Aborigines, fire and Australian ecosystems. *Mankind* **13**: 237–51.

Huggett, R.J. 1997: *Environmental change: the evolving ecosphere*. London, Routledge.

Hughes, C. 1995: One land, two laws: Aboriginal fire management. *Environmental and Planning Law Journal* **12**: 37–49.

Huntley, B. 1994: Rapid early-Holocene migration and high abundance of hazel (*Corylus avellana* L.): alternative hypotheses. In Chambers, F.M. (ed.), *Climate Change and Human Impact on the Landscape*. London, Chapman & Hall: 205–16.

Huntley, B., Cramer, W., Morgan, A.V., Prentice, H.C. and **Allen, J.R.M.** 1997: Predicting the response of terrestrial biota to future environmental changes. In Huntley, B., Cramer, W., Morgan, A.V., Prentice, H.C. and Allen, J.R.M. (eds), *Past and future rapid environmental changes: the spatial and evolutionary responses of terrestrial biota*. Berlin, Springer: 487–504.

Hynes, R.A. and **Chase, A.** 1982: Plants, sites and domi-culture: Aboriginal influence upon plant communities in Cape York Peninsula. *Archaeology in Oceania* **17**: 38–50.

Ikawa-Smith, F. 1999: Constructions of national identity and origins in East Asia: a comparative perspective. *Antiquity* **73**: 626–9.

Jackson, P. 1989: *Maps of meaning: an introduction to cultural geography*. London, Unwin Hyman.

Jackson, P., Cosgrove, D., Duncan, J. and **Duncan, N.** 1996: A debate on Mitchell, D. 1995 'There's no such thing as culture'. *Transactions of the Institute of British Geographers* NS **21**: 572–82.

Jackson, S.T., Overpeck, J.T., Webb, T., Keattch, S.E. and **Anderson, K.H.** 1997: Mapped plant-macrofossil and pollen records of late Quaternary vegetation change in eastern North America. *Quaternary Science Reviews* **16**: 1–70.

Jacobs, J.M. and **Gale, F.** 1994: *Tourism and the protection of Aboriginal cultural sites*. Canberra, Australian Government Publishing Service.

Jasanoff, S. and **Wynne, B.** 1998: Science and decision-making. In Rayner, S. and Malone, E.L. (eds), *Human choice and climate change*, vol. 1, *The societal framework*. Columbus, OH, Battelle Press: 1–88.

Jian-Hua, L. and **Hills, P.** 1997: Marine protected areas and local coastal conservation and management in Hong Kong. *Local Environment* **2**: 275–97.

Jiantian, G. 1998: Conservation of plant diversity in China: achievements, prospects and concerns. *Biological Conservation* **85**: 321–7.

Johnston, R.J., Gregory, D. and **Smith, D.M.** (eds) 1994: *The dictionary of human geography*. Oxford, Blackwell.

Jolly, D., Taylor, D., Marchant, R., Hamilton, A., Bonnefille, R., Buchet, G. and **Riollet, B.** 1997: Vegetation dynamics in central Africa since 18,000 yr BP: pollen records from the interlacustrine highlands of Burundi, Rwanda and western Uganda. *Journal of Biogeography* **24**: 495–512.

Jones, M. and **Daugstad, K.** 1997: Usages of the 'cultural landscape' concept in Norwegian and Nordic landscape administration. *Landscape Research* **22**(3): 267–81.

Kealhofer, L. 1999: Creating social identity in the landscape: Tidewater, Virginia, 1600–1750. In Ashmore, W. and Knapp, A.B. (eds), *Archaeologies of Landscape*. Oxford, Blackwell: 58–82.

Keating, P.J. 1993: Prime Minister's Address to the Nation. In Goot, M. and Rowse, T. (eds), *Make a better offer: the politics of Mabo*. Sydney, Pluto Press: 235–8.

Keenan, S. and **Davidson, P.** 1997: The iconography of the Globe. In Mulryne, J.R. and Shewring, M. (eds), *Shakespeare's Globe rebuilt*. Cambridge, Cambridge University Press: 147–56.

Kern, K. 1992: Rehabilitation of streams in south-west Germany. In Boon, P.J., Calow, P. and Petts, G.E. (eds), *River conservation and management*. Chichester, John Wiley: 321–36.

Kershaw, A.P. 1986: The last two glacial–interglacial cycles from northeastern Australia: implications for climate change and Aboriginal burning. *Nature* **322**: 47–9.

Kershaw, A.P., Clark, J.S., Gill, A.M. and **D'Costa, D.M.** (in press). A history of fire in Australia. In Bradstock, R., Williams, J. and Gill, A.M. (eds), *Flammable Australia: the fire regimes and biodiversity of a continent*. Cambridge, Cambridge University Press.

Kershaw, A.P., McKenzie, G.M. and **McMinn, A.** 1993: A Quaternary vegetation history of northeastern Queensland from pollen analysis of ODP site 820. *Proceedings of the Ocean Drilling Program, Scientific Results* **133**: 107–14.

Kirby, V.G. 1993: Landscape, heritage and identity: stories from the West Coast. In Hall, C.M. and McArthur, S. (eds), *Heritage management in New Zealand and Australia*. Auckland, Oxford University Press: 119–29.

Kirch, P.V. 1984: *The evolution of the Polynesian chiefdoms*. Cambridge, Cambridge University Press.

Kirch, P.V. and **Ellison, J.** 1994: Palaeoenvironmental evidence for human colonization of remote oceanic islands. *Antiquity* **68**: 310–21.

Kirikiri, R. and **Nugent, G.** 1995: Harvesting of New Zealand native birds by Maori. In Grigg, G.C., Hale, P.T. and Lunney, D. (eds), *Conservation through sustainable use of wildlife*. Brisbane, Centre for Conservation Biology, University of Queensland: 54–9.

Klomp, N. and **Lunt, I.** (eds) 1997: *Frontiers in ecology: building the links*. Oxford, Elsevier Science.

Koff, T., Punning, J.-M. and **Yli-Halla, M.** 1998: Human impact on a paludified landscape in northern Estonia. *Landscape and Urban Planning* **41**: 263–72.

Kondolf, G.M. 1995: Five elements for effective evaluation of stream restoration. *Restoration Ecology* **3**: 133–6.

Kuklick, H. 1993: *The savage within: the social history of British anthropology, 1885–1945*. Cambridge, Cambridge University Press.

Kuper, A. 1988: *The invention of primitive society: transformations of an illusion*. London, Routledge.

Langton, M. 1995–6: The European construction of wilderness. *Wilderness News* **143**: 16–17.

Langton, M. 1996: Art, wilderness and *terra nullius*. *Conference Papers and Resolutions*. Ecopolitics IX. Darwin, Northern Land Council: 11–24.

Langton, M. 1998: *Burning questions: emerging environmental issues for indigenous peoples in northern Australia*. Darwin, Centre for Indigenous Natural and Cultural Resource Management.

Langton, M., Epworth, D. and **Sinnamon, V.** 1999: *Indigenous social, economic and cultural issues in land, water and biodiversity conservation: a scoping study for WWF Australia*, vol. 1. Darwin, Centre for Indigenous Natural and Cultural Resource Management.

Ley, D. 1985: Cultural/humanistic geography. *Progress in Human Geography* **9**: 415–23.

Leyden, B.W., Brenner, M. and **Dahlin, B.H.** 1998: Cultural and climatic history of Coba, a lowland Maya city in Quintana Roo, Mexico. *Quaternary Research* **49**: 111–22.

Lourandos, H. 1983: Intensification: a late Pleistocene–Holocene archaeological sequence from south-western Victoria. *Archaeology in Oceania* **18**: 81–94.

Lourandos, H. 1993: Hunter-gatherer cultural dynamics: long- and short-term trends in Australian prehistory. *Journal of Archaeological Research* **1**(1): 67–88.

Lowe, D.J., McFadgen, B.G., Higham, T.F.G., Hogg, A.G., Froggatt, P.C. and **Nairn, I.A.** 1998: Radiocarbon age of the Kaharoa Tephra, a key marker for late-Holocene stratigraphy and archaeology in New Zealand. *The Holocene* **8**: 487–95.

Lowenthal, D. 1978: Australian images: the unique present, the mythical past. In Quartermaine, P. (ed.), *Readings in Australian Arts*. Exeter, University of Exeter: 84–94.

Lowenthal, D. 1996: *Possessed by the past: the heritage crusade and the spoils of history*. New York, The Free Press.

Lye, T.-P. 1998: *Hep*: the significance of forest to the emergence of Batek knowledge in Pahang, Malaysia. Paper presented at CHAGS 8, Osaka, Japan.

Maley, J. and **Brenac, P.** 1998: Vegetation dynamics, palaeoenvironments and climatic changes in the forests of western Cameroon during the last 28,000 years BP. *Review of Palaeobotany and Palynology* **99**: 157–87.

Maloney, B.K. 1994: The prospects and problems of using palynology to trace the origins of tropical agriculture: the case of Southeast Asia. In Hather, J. (ed.), *Tropical archaeobotany: applications and new developments*. London, Routledge: 139–71.

Mannion, A.M. 1991: *Global environmental change: a natural and cultural enviromental history*. Harlow, Longman.

Marsh, G.P. 1965 [1864]: *Man and nature*. Cambridge, MA, The Belknap Press of Harvard University Press.

Mazel, A.D. 1992: Changing fortunes: 150 years of San hunter-gatherer history in the Natal Drakensberg, South Africa. *Antiquity* **66**: 758–67.

McAndrews, J.H. 1988: Human disturbance of North American forests and grasslands: the fossil pollen record. In Huntley, B. and Webb, T. (eds), *Vegetation history*. Dordrecht, Kluwer: 673–98.

McDonnell, M.J. and **Pickett, S.T.A.** (eds) 1993a: *Humans as components of ecosystems: the ecology of subtle human effects and populated areas*. New York, Springer-Verlag.

McDonnell, M.J. and **Pickett, S.T.A.** 1993b: Introduction: scope and need for an ecology of subtle human effects and populated areas. In McDonnell, M.J. and Pickett, S.T.A. (eds), *Humans as components of ecosystems: the ecology of subtle human effects and populated areas*. New York, Springer-Verlag: 1–6.

McGlone, M. 1999: The role of history in the creation of a vision of future landscapes. Paper presented to Fenner Conference, 'Visions of Future Landscapes', Canberra, 2–5 May.

McGlone, M.S. 1983: Polynesian deforestation of New Zealand: a preliminary synthesis. *Archaeology in Oceania* **18**: 11–25.

McGlone, M.S. and **Wilmshurst, J.M.** 1999: Dating initial Maori environmental impact in New Zealand. *Quaternary International* **59**: 5–16.

McNiven, I. and **Russell, L.** 1995: Place with a past: reconciling wilderness and the Aboriginal past in World Heritage areas. *Royal Historical Society of Queensland Journal* **15**(11): 505–19.

Meadows, M.E. 1999: Biogeography: changing places, changing times. *Progress in Physical Geography* **23**: 257–70.

Mearns, L. 1994: To continue the Dreaming: Aboriginal women's traditional responsibilities in a transformed world. In Burch, E.S. and Ellanna, L.J. (eds), *Key issues in hunter-gatherer research*. Oxford, Berg: 263–88.

Meiklejohn, K.I. 1995: The deterioration and preservation of rock art in the KwaZulu–Natal Clarens Formation: a geomorphological perspective. *Pictogram* **8**(1): 1–13.

Meinig, D.W. 1962: *On the margins of the good earth*. Australia, Rigby.

Merchant, C. 1980: *The death of nature: women, ecology and the scientific revolution*. San Francisco, Harper & Row.

Merchant, C. 1989: *Ecological revolutions: nature, gender, and science in New England*. Chapel Hill, University of North Carolina Press.

Meyer, W.B. 1996: *Human impact on the Earth*. Cambridge, Cambridge University Press.

Meyer, W.B., Butzer, K.W., Downing, T.E., Turner, B.L., Wenzel, G.W. and **Wescoat, J.L.** 1998: Reasoning by analogy. In Rayner, S. and Malone, E.L. (eds), *Human choice and climate change*, vol. 3, *The tools for policy analysis*. Columbus, OH, Battelle Press: 217–290.

Mitchell, D. 1995: There's no such thing as culture: towards a reconceptualization of the idea of culture in geography. *Transactions of the Institute of British Geographers* NS **20**: 102–16.

Mitchell, N.J. 1995: Cultural landscapes in the United States. In von Droste, B., Plachter, H. and Rossler, M. (eds), *Cultural landscapes of universal value: components of a global strategy*. New York and Jena, Gustav Fischer Verlag: 234–51.

Moller, H. 1996: Customary use of indigenous wildlife: towards a bicultural approach to conserving New Zealand's biodiversity. In McFadgen, B. and Simpson, P. (eds), *Biodiversity: papers from a seminar series on biodiversity, hosted by Science and Research Division, Dept. of Conservation, Wellington*: 89–125.

Moore, P.D. 1994: The origin of blanket mire, revisited. In Chambers, F.M. (ed.), *Climate change and human impact on the landscape*. London, Chapman & Hall: 217–24.

Moore, P.D., Chaloner, B. and **Stott, P.** 1996: *Global environmental change*. Oxford, Blackwell.

Moorehead, A. 1963: *Cooper's Creek*. London, Hamish Hamilton.

Muir, R. 1998: Reading the landscape, rejecting the present. *Landscape Research* **23**(1): 71–82.

Mulk, I.-M. and **Bayliss-Smith, T.** 1999: The representation of Sami cultural identity in the cultural landscapes of northern Sweden: the use and misuse of archaeological knowledge. In Ucko, P.J. and Layton, R. (eds), *The archaeology and anthropology of landscape: shaping your landscape*. London, Routledge: 358–95.

Mulryne, J.R. and **Shewring, M.** 1997a: The once and future Globe. In Mulryne, J.R. and Shewring, M. (eds), *Shakespeare's Globe rebuilt*. Cambridge, Cambridge University Press: 15–25.

Mulryne, J.R. and **Shewring, M.** (eds) 1997b: *Shakespeare's Globe rebuilt*. Cambridge, Cambridge University Press.

Mulvaney, D.J. and **Kamminga, J.** 1999: *Prehistory of Australia*. Sydney, Allen & Unwin.

Nash, R. 1967: *Wilderness and the American mind*. New Haven, CT, Yale University Press.

Naughton-Treves, L. 1998: Predicting patterns of crop damage by wildlife around Kibale National Park, Uganda. *Conservation Biology* **12**: 156–68.

Newmark, W.D. 1996: Insularization of Tanzanian parks and the local extinction of large mammals. *Conservation Biology* **10**: 1549–56.

Newnham, R.M., Lowe, D.J. and **Matthews, B.W.** 1998a: A late-Holocene and prehistoric record of environmental change from Lake Waikaremoana, New Zealand. *The Holocene* **8**: 443–54.

Newnham, R.M., Lowe, D.J., McGlone, M.S., Wilmshurst, J.M. and **Higham, T.F.G.** 1998b: The Kaharoa Tephra as a critical datum for earliest human impact in northern New Zealand. *Journal of Archaeological Science* **25**: 533–44.

Nunn, P.D. 1994: Beyond the naive lands: human history and environmental change in the Pacific Basin. In Waddell, E. and Nunn, P.D. (eds), *The margin fades: geographical itineraries in a world of islands*. Suva, Institute of Pacific Studies, University of the South Pacific: 5–27.

Nunn, P.D. 1999: *Environmental change in the Pacific Basin: chronologies, causes, consequences*. Chichester, John Wiley.

O'Connell, J.F. and **Allen, J.** 1998: When did humans first arrive in Greater Australia and why is it important to know? *Evolutionary Anthropology* **6**: 132–46.

Ogden, J., Basher, L. and **McGlone, M.** 1998: Fire, forest regeneration and links with early human habitation: evidence from New Zealand. *Annals of Botany* **81**: 687–96.

Oldfield, F. 1994: Forward to the past: changing approaches to Quaternary palaeoecology. In Chambers, F.M. (ed.), *Climate change and human impact on the landscape*. London, Chapman & Hall: 13–22.

Olwig, K.R. 1996: Reinventing common nature: Yosemite and Mount Rushmore – a meandering tale of a double nature. In Cronon, W. (ed.), *Uncommon ground: rethinking the human place in nature*. New York, W.W. Norton: 379–408.

O'Riordan, T. 1998: Sustainability for survival in South Africa. *Global Environmental Change* **8**: 99–108.

O'Riordan, T., Cooper, C.L., Jordan, A. *et al.* 1998: Institutional frameworks for political action. In Rayner, S. and Malone, E.L. (eds), *Human choice and climate change*, vol. 1, *The societal framework*. Columbus, Battelle Press: 345–39.

Osborne, L.L., Bailey, P.B., Higler, L.W.G., Statzner, B., Triska, F. and **Moth Iversen, T.** 1993: Restoration of lowland streams: an introduction. *Freshwater Biology* **29**: 187–94.

Pai, H.I. 1999: Nationalism and preserving Korea's buried past: the Office of Cultural Properties and archaeological heritage management in South Korea. *Antiquity* **73**: 619–25.

Pak, Y. 1999: Contested ethnicities and ancient homelands in northeast Chinese archaeology: the case of Koguryo and Puyo archaeology. *Antiquity* **73**: 613–18.

Pardoe, C. 1988: The cemetery as symbol. The distribution of prehistoric Aboriginal burial grounds in southeastern Australia. *Archaeology in Oceania* **23**: 1–16.

Parker, V.T. 1997: The scale of successional models and restoration objectives. *Restoration Ecology* **5**: 301–6.

Parker, V.T. and **Pickett, S.T.A.** 1997: Restoration as an ecosystem process: implications of the modern ecological paradigm. In Urbanska, K.M., Webb, N.R. and Edwards, P.J. (eds), *Restoration ecology and sustainable development*. Cambridge, Cambridge University Press: 17–32.

Pawson, E. 1992: Two New Zealands: Maori and European. In Anderson, K. and Gale, F. (eds), *Inventing places: studies in cultural geography*. Melbourne, Longman Cheshire: 15–36.

Peacock, R.J., Williams, J.E. and **Franklin, J.F.** 1997: Disturbance ecology of forested ecosystems: implications for sustainable management. In Klomp, N. and Lunt, I. (eds), *Frontiers in Ecology: Building the Links*. Oxford, Elsevier Science: 67–78.

Pearsall, D.M. 1994: Investigating New World tropical agriculture: contributions from phytolith analysis. In Hather, J. (ed.), *Tropical archaeobotany: applications and new developments*. London, Routledge: 115–38.

Phillips, A. 1995: Cultural landscapes: an IUCN perspective. In von Droste, B., Plachter, H. and Rossler, M. (eds), *Cultural landscapes of universal value: components of a global strategy*. New York and Jena, Gustav Fisher Verlag: 380–92.

Pickett, S.T.A. and **Parker, V.T.** 1994: Avoiding the old pitfalls: opportunities in a new discipline. *Restoration Ecology* 2: 75–9.

Pickett, S.T.A. and **White, P.S.** (eds) 1985a: *The ecology of natural disturbance and patch dynamics*. Orlando, FL, Academic Press.

Pickett, S.T.A. and **White, P.S.** 1985b: Patch dynamics: a synthesis. In Pickett, S.T.A. and White, P.S. (eds), *The ecology of natural disturbance and patch dynamics*. Orlando, FL, Academic Press: 371–84.

Povinelli, E.A. 1993: *Labor's lot: the power, history, and culture of Aboriginal action*. Chicago, University of Chicago Press.

Powell, J.M. 1977: *Mirrors of the New World: images and image-makers in the settlement process*. Folkestone, Dawson.

Powell, J.M. 1988a: *An historical geography of modern Australia: the restive fringe*. Cambridge, Cambridge University Press.

Powell, J.M. 1988b: Protracted reconciliation: society and the environment. In MacLeod, R. (ed.), *The Commonwealth of Science: ANZAAS and the scientific enterprise in Australasia, 1888–1988*. Melbourne, Oxford University Press: 249–71.

Price, M. and **Lewis, M.** 1993: The reinvention of cultural geography. *Annals of the Association of American Geographers* 83: 1–17.

Proctor, J.D. 1998: The meaning of global environmental change: retheorizing culture in human dimensions research. *Global Environmental Change* 8: 227–48.

Proctor, J.D. and **Pincetl, S.** 1996: Nature and the reproduction of endangered space: the spotted owl in the Pacific Northwest and southern California. *Environment and Planning D: Society and Space* 14: 683–708.

Punning, J.-M. (ed.) 1994: *The influence of natural and anthropogenic factors on the development of landscapes: the results of a comprehensive study in NE Estonia*. Tallinn, Institute of Ecology, Estonian Academy of Sciences.

Punning, J.-M., Boyle, J.F., Alliksaaw, T., Tann, R. and **Varvas, M.** 1997a: Human impact on the history of Lake Nommejarv, NE Estonia: a geochemical and palaeobotanical study. *The Holocene* 7: 91–9.

Punning, J.-M., Koff, T., Tann, R. and **Lukki, T.** 1997b: The sensitivity and adaptation of ecosystems to the disturbances: a case study in northeastern Estonia. *Mitigation and Adaptation Strategies for Global Change* 2: 1–17.

Ramos, A. 1994: From Eden to limbo: the construction of indigenism in Brazil. In Bond, G.C. and Gilliam, A. (eds),

Social construction of the past: representation as power. London, Routledge: 74–88.

Rao, N. 1994: Interpreting silences: symbol and history in the case of Ram Janmabhoomi / Babri Masjid. In Bond, C. and Gilliam, A. (eds), *Social construction of the past: representation as power*. London, Routledge: 154–66.

Ratzel, F. 1895–6: Die deutsche Landschaft. *Halbmonatshefte der deutschen Rundschau* 4: 407–28.

Rayner, S. and **Malone, E.L.** (eds) 1998a: *Human choice and climate change*. 4 vols. Columbus, OH, Battelle Press.

Rayner, S. and **Malone, E.L.** 1998b: Social science insights into climate change. In Rayner, S. and Malone, E.L. (eds), *Human choice and climate change*, vol. 4, 'What have we learned?' Columbus, OH, Battelle Press: 71–108.

Richardson, J.A. 1998: Wildlife utilization and biodiversity conservation in Namibia: conflicting or complementary objectives? *Biodiversity and Conservation* 7: 549–59.

Robbins, R. 1998: Paper forests: imagining and deploying exogenous ecologies in arid India. *Geoforum* 29(1): 69–86.

Roberts, M., Norman, W., Minhinnick, N., Wihongi, D. and **Kirkwood, C.** 1995: Kaitiakitanga: Maori perspectives on conservation. *Pacific Conservation Biology* 2: 7–20.

Roberts, N. 1989: *The Holocene: an environmental history*. Oxford, Basil Blackwell.

Roberts, R., Bird, M., Olley, J. *et al.* 1998: Optical and radiocarbon dating at Jinmium rock shelter in northern Australia. *Nature* 393: 358–62.

Roberts, R.G., Jones, R. and **Smith, M.A.** 1990: Thermoluminescence dating of a 50,000-year-old human occupation site in northern Australia. *Nature* 345: 153–6.

Roberts, R.G., Jones, R., Spooner, N.A., Head, M.J., Murray, A.S. and **Smith, M.A.** 1994: The human colonization of Australia: optical dates of 53,000 and 60,000 years bracket human arrival at Deaf Adder Gorge, Northern Territory. *Quaternary Science Reviews* 13: 575–83.

Rodbell, D.T., Seltzer, G.O., Anderson, D.M., Abbott, M.B., Enfield, D.B. and **Newman, J.H.** 1999: An ~15,000-year record of El Niño-driven alluviation in southwestern Ecuador. *Science* 283: 516–20.

Rose, B. 1995: *Land management issues: attitudes and perceptions amongst Aboriginal people of central Australia*. Alice Springs, Central Land Council.

Rossignol, J. and **Wandsnider, L.** (eds) 1992: *Space, time and archaeological landscapes*. New York, Plenum Press.

Rossler, M. 1995: UNESCO and cultural landscape protection. In van Droste, B., Plachter, H. and Rossler, M. (eds), *Cultural landscapes of universal value – components of a global strategy*. New York and Jena, Gustav Fischer Verlag: 42–9.

Rowlands, M. 1994: The politics of identity in archaeology. In Bond, G.C. and Gilliam, A. (eds), *Social construction of the past: representation as power*. London, Routledge: 129–43.

Rowntree, L.B. 1996: The cultural landscape concept in American human geography. In Earle, C., Mathewson, K. and Kenzer, M.S. (eds), *Concepts in human geography*. Lanham, MD, Rowman & Littlefield: 127–60.

Russell-Smith, J. and **Dunlop, C.** 1987: The status of monsoon vine forests in the Northern Territory: a perspective. In Werren, G. and Kershaw, A.P. (eds), *The rainforest legacy*. Canberra, Australian Heritage Commission Special Australian Heritage Publication Series 7: 227–88.

Saberwal, V.K. and **Kothari, A.** 1996: The human dimension in conservation biology curricula in developing countries. *Conservation Biology* **10**(5): 1328–31.

Safford, R.J. 1997: A survey of the occurrence of native vegetation remnants on Mauritius in 1993. *Biological Conservation* **80**: 181–8.

Sala, O.E., Chapin, F.S., Gardner, R.H., Lauenroth, W.K., Mooney, H.A. and **Ramakrishnan, P.S.** 1999: Global change, biodiversity and ecological complexity. In Walker, B., Steffen, W., Canadell, J. and Ingram, J. (eds), *The terrestrial biosphere and global change: implications for natural and managed ecosystems.* Cambridge, Cambridge University Press: 304–28.

Saleh, M.A.E. 1997: Toward a sustainable land management of vernacular landscape in the highlands of south-western Saudi Arabia: indigenous and statutory experiences. *Landscape Research* **22**: 283–302.

Sandweiss, D.H., Richardson, J.B., Reitz, E.J., Rollins, H.B. and **Maasch, K.A.** 1996: Geoarchaeological evidence from Peru for a 5000 years B.P. onset of El Niño. *Science* **273**: 1531–3.

Sauer, C. 1965 [1925]: The morphology of landscape. In *Land and life: a selection from the writings of Carl Ortwin Sauer,* ed. Leighly, J. Berkeley, University of California Press: 315–50.

Schein, R.H. 1997: The place of landscape: a conceptual framework for interpreting an American scene. *Annals of the Association of American Geographers* **87**(4): 660–80.

Schmidt, P.R. and **Patterson, T.C.** (eds) 1995: *Making alternative histories: the practice of archaeology and history in non-Western settings.* Santa Fe, NM, School of American Research Press.

Shackley, S. and **Wynne, B.** 1995: Integrating knowledges for climate change: pyramids, nets and uncertainties. *Global Environmental Change* **5**: 113–26.

Sherratt, A. 1995: Instruments of conversion? The role of megaliths in the Mesolithic/Neolithic transition in north-west Europe. *Oxford Journal of Archaeology* **14**: 245–60.

Sibley, D. 1992: Outsiders in society and space. In Anderson, K. and Gale, F. (eds), *Inventing places: studies in cultural geography.* Melbourne, Longman Cheshire: 107–22.

Siegfried, W.R., Benn, G.A. and **Gelderblom, C.M.** 1998: Regional assessment and conservation implications of landscape characteristics of African national parks. *Biological Conservation* **84**: 131–40.

Simmons, I.G. 1989: *Changing the face of the Earth: culture, environment, history.* Oxford, Basil Blackwell.

Simmons, I.G. 1993a: Human societies and environmental change: the long view. In Johnston, R.J. (ed.), *The challenge for geography. A changing world: a changing discipline.* Oxford, Blackwell: 100–16.

Simmons, I.G. 1993b: *Interpreting nature: cultural constructions of the environment.* London, Routledge.

Simmons, I.G. 1994: Vegetation change during the Mesolithic in the British Isles: some amplifications. In Chambers, F.M. (ed.), *Climate change and human impact on the landscape.* London, Chapman & Hall: 109–18.

Simpson, I.A., Parsisson, D., Hanley, N. and **Bullock, C.H.** 1997: Envisioning future landscapes in the Environmentally Sensitive Areas of Scotland. *Transactions of the Institute of British Geographers* NS **22**: 307–20.

Singh, G., Kershaw, P. and **Clark, R.** 1981: Quaternary vegetation and fire history in Australia. In Gill, A.M., Groves, R.H. and Noble, I.R. (eds), *Fire and the*

Australian Biota. Canberra, Australian Academy of Science: 23–54.

Skira, I. 1996: Aboriginal people and muttonbirding in Tasmania. In Bomford, M. and Caughley, J. (eds), *Sustainable use of wildlife by Aboriginal peoples and Torres Strait Islanders.* Canberra, Australian Government Publishing Service: 167–75.

Sluyter, A. 1997: On 'Buried epistemologies: the politics of nature in (post)colonial British Columbia'. On excavating and burying epistemologies. *Annals of the Association of American Geographers* **87**(4): 700–2.

Smyth, D. and **Sutherland, J.** 1996: *Indigenous protected areas: conservation partnerships with indigenous landholders.* 2 vols. Canberra, Environment Australia.

Soulé, M.E. 1995: The social siege of nature. In Soulé, M.E. and Lease, G. (eds), *Reinventing nature? Responses to postmodern deconstruction.* Washington, DC, Island Press: 137–71.

Soulé, M.E. and **Lease, G.** (eds) 1995: *Reinventing nature? Responses to postmodern deconstruction.* Washington, DC, Island Press.

Spiegel, A.D. 1994: Struggling with tradition in South Africa: the multivocality of images of the past. In Bond, G.C. and Gilliam, A. (eds), *Social construction of the past: representation as power.* London, Routledge: 185–202.

Spirn, A.W. 1996: Constructing nature: the legacy of Frederick Law Olmsted. In Cronon, W. (ed.), *Uncommon ground: rethinking the human place in nature.* New York, W.W. Norton: 91–113.

Spriggs, M. and **Anderson, A.** 1993: Late colonization of east Polynesia. *Antiquity* **67**: 200–17.

Storli, I. 1996: On the historiography of Sámi reindeer pastoralism. *Acta Borealia* **13**: 81–115.

Stott, P. 1998: Biogeography and ecology in crisis: the urgent need for a new metalanguage. *Journal of Biogeography* **25**: 1–2.

Sugden, D.E. 1996: The East Antarctic ice sheet: unstable ice or unstable ideas? *Transactions of the Institute of British Geographers* NS **21**: 443–54.

Sullivan, S. 1996: Towards a non-equilibrium ecology: perspectives from an arid land. *Journal of Biogeography* **23**: 1–5.

Sultan, R. 1991: A voice in the wilderness? Aboriginal perspectives on conservation. *Habitat* **19**: 1.

Sutton, D.G. 1987: A paradigmatic shift in Polynesian prehistory: implications for New Zealand. *New Zealand Journal of Archaeology* **9**: 135–55.

Suzuki, D. and **Knudtson, P.** 1992: *Wisdom of the elders: sacred native stories of nature.* New York, Bantam Books.

Sykes, M.T. 1997: The biogeographic consequences of forecast changes in the global environment: individual species' potential range changes. In Huntley, B., Cramer, W., Morgan, A.V., Prentice, H.C. and Allan, J.R.M. (eds), *Past and future rapid environmental changes: the spatial and evolutionary responses of terrestrial biota.* Berlin, Springer: 427–40.

Taçon, P.S.C. 1991: The power of stone: symbolic aspects of stone use and tool development in western Arnhem Land, Australia. *Antiquity* **65**: 192–207.

Taçon, P.S.C. 1994: Socialising landscapes: the long-term implications of signs, symbols and marks on the land. *Archaeology in Oceania* **29**: 117–29.

Taçon, P.S.C., Fullagar, R., Ouzman, S. and **Mulvaney, K.** 1997: Cupule engravings from Jinmium-Granilpi (northern Australia) and beyond: exploration of a wide-spread and enigmatic class of rock markings. *Antiquity* **71**: 942–65.

Taiepa, T., Lyver, P., Horsley, P., Davis, J., Bragg, M. and **Moller, H.** 1996: Collaborative management of New Zealand's conservation estate by Maori and Pakeha. Unpublished manuscript.

Taylor, D. 1993: Environmental change in montane south-west Uganda: a pollen record for the Holocene from Ahakagyezi Swamp. *The Holocene* **3**: 324–32.

Taylor, D., Robertshaw, P. and **Marchant, R.A.** 1999: Determining the impact of humans on sedimentary sequences in the interlacustrine region of central Africa. Paper given at the 15th International INQUA Congress, Durban, South Africa.

Thackway, R., Szabo, S. and **Smyth, D.** 1996: Indigenous protected areas: a new concept in biodiversity conservation. *Biodiversity: broadening the debate 4*. Canberra, Australian Nature Conservation Agency: 18–34.

Thomas, J. 1991: *Rethinking the Neolithic*. Cambridge, Cambridge University Press.

Thomas, J. 1996: *Time, culture and identity: an interpretive archaeology*. London, Routledge.

Thomas, N. 1994: *Colonialism's culture: anthropology, travel and government*. Cambridge, Polity Press.

Thomas, R.C., Kirby, K.J. and **Reid, C.M.** 1997: The conservation of a fragmented ecosystem within a cultural landscape: the case of ancient woodland in England. *Biological Conservation* **82**: 243–52.

Thomas, W.L. (ed.) 1956: *Man's role in changing the face of the Earth*. Chicago: University of Chicago Press.

Thompson, M. and **Rayner, S.** 1998: Cultural discourses. In Rayner, S. and Malone, E.L. (eds), *Human choice and climate change*, vol. 1, *The societal framework*. Columbus, Battelle Press: 265–344.

Thorne, A., Grun, R., Mortimer, G. *et al.* 1999: Australia's oldest human remains: age of the Lake Mungo 3 skeleton. *Journal of Human Evolution* **36**: 591–612.

Titchen, S. 1996: Including cultural landscapes on the World Heritage List. *World Heritage Review* **2**: 34–9.

Tuan, Y.-F. 1968: Discrepancies between environmental attitude and behaviour: examples from Europe and China. *Canadian Geographer* **12**: 176–91.

Tuan, Y.-F. 1971: Man and nature. Association of American Geographers Resource Paper 10.

Turner, B.L., Clark, W.C., Kates, R.W., Richards, J.F., Mathews, J.T. and **Meyer, W.B.** (eds) 1990: *The Earth as transformed by human action*. Cambridge, Cambridge University Press.

Ucko, P.J. and **Layton, R.** (eds) 1999: *The archaeology and anthropology of landscape: shaping your landscape*. London, Routledge.

UNESCO 1995: *Report on the Asia-Pacific regional workshop on associative cultural landscapes*. Sydney.

UNESCO 1997: World Heritage Committee, 20th session, Nerida, Mexico, 2–7 December 1996.

UNESCO 1999: www.unesco.org/whc/exhibits/afr_rev/africa-b.htm.

Van Gijn, A. and **Zvelebil, M.E.** 1997: Ideology and social structure of Stone Age communities in Europe. *Analecta Praehistorica Leidensia* **29**.

Varvas, M. and **Punning, J.-M.** 1993: Use of the ^{210}Pb method in studies of the development and human-impact history of some Estonian lakes. *The Holocene* **3**: 34–44.

Villers-Ruiz, L. and **Trejo-Vazquez, I.** 1998: Climate change on Mexican forests and natural protected areas. *Global Environmental Change* **8**: 141–57.

Vitousek, P.M., Mooney, H.A., Lubchenco, J. and **Melillo, J.M.** 1997: Human domination of Earth's ecosystems. *Science* **277**: 494–9.

Wahl, E.J. 1999: Rock paintings in Giant's Castle Game Reserve, uKhahlamba (Drakensberg), KwaZulu–Natal, South Africa. 15th International Inqua Conference Field Guide. Amafa aKwaZulu-Natal, Pietermaritzburg.

Waitt, G. 1997: Selling paradise and adventure: representations of landscape in the tourist advertising of Australia. *Australian Geographical Studies* **35**(1): 47–60.

Walker, B.H. and **Steffen, W.L.** 1999: The nature of global change. In Walker, B., Steffen, W., Canadell, J. and Ingram, J. (eds), *The terrestrial biosphere and global change: implications for natural and managed ecosystems*. Cambridge, Cambridge University Press: 1–18.

Walker, B.H., Steffen, W.L. and **Langridge, J.** 1999: Interactive and integrated efects of global change on terrestrial ecosystems. In Walker, B., Steffen, W., Canadell, J. and Ingram, J. (eds), *The terrestrial biosphere and global change*. Cambridge, Cambridge University Press: 329–75.

Walker, D. and **Singh, G.** 1994: Earliest palynological records of human impact on the world's vegetation. In Chamber, F.M. (eds), *Climate change and human impact on the landscape*. London, Chapman & Hall: 101–8.

Webb, G., Missi, C. and **Cleary, M.** 1996: Sustainable use of crocodiles by Aboriginal people in the Northern Territory. In Bomford, M. and Caughley, J. (eds), *Sustainable use of wildlife by Aboriginal peoples and Torres Strait Islanders*. Canberra, Australian Government Publishing Service: 176–86.

West, R.G. 1956: The Quaternary deposits at Hoxne, Suffok. *Philosophical Transactions of the Royal Society of London B* **239**: 265–356.

West, R.G. and **McBurney, C.M.B.** 1954: The Quaternary deposits at Hoxne, Suffok, and their archaeology. *Proceedings of the Prehistoric Society* **20**: 131–54.

Whatmore, S. and **Thorne, L.** 1998: Wild(er)ness: reconfiguring the geographies of wildlife. *Transactions of the Institute of British Geographers* NS **23**: 435–54.

White, P.S. and **Pickett, S.T.A.** 1985: Natural disturbance and patch dynamics: an introduction. In Pickett, S.T.A. and White, P.S. (eds), *The ecology of natural disturbance and patch dynamics*. Orlando, FL, Academic Press: 3–13.

Whittle, A. 1996: *Europe in the Neolithic: the creation of new worlds*. Cambridge, Cambridge University Press.

Willems-Braun, B. 1997: Reply: on cultural politics, Sauer, and the politics of citation. *Annals of the Association of American Geographers* **87**: 703–8.

Williams, M., Dunkerley, D., De Deckker, P., Kershaw, P. and **Chappell, J.** 1998: *Quaternary environments*. London, Arnold.

Williams, R. 1980: *Problems in materialism and culture*. London, Verso.

Willis, K.J. and **Bennett, K.D.** 1994: The Neolithic transition – fact or fiction? Palaeoecological evidence from the Balkans. *The Holocene* **4**: 326–30.

Wilmshurst, J. and **McGlone, M.** 1998: Recent South Island sphagnum bogs: anthropogenic or climatic origins? Paper presented to New Zealand Ecological Society, Dunedin.

Wiltshire, P.E.J. and **Edwards, K.J.** 1994: Mesolithic, early Neolithic, and later prehistoric impacts on vegetation at a riverine site in Derbyshire, England. In Chambers, F.M. (ed.), *Climate change and human impact on the landscape.* London, Chapman & Hall: 157–68.

Worster, D. 1985: *Rivers of empire: water, aridity, and the growth of the American West.* New York, Oxford University Press.

Wu, J. and **Loucks, O.L.** 1995: From balance of nature to hierarchical patch dynamics: a paradigm shift in ecology. *Quarterly Review of Biology* **70**: 439–66.

Wylie, A. 1995: Alternative histories: epistemic disunity and political integrity. In Schmidt, P.R. and Patterson, T.C. (eds), *Making alternative histories: the practice of archaeology and history in non-Western settings.* Santa Fe, NM, School of American Research Press: 255–72.

Wynne, B. 1994: Scientific knowledge and the global environment. In Redclift, M. and Benton, T. (eds), *Social theory and the global environment.* London, Routledge: 169–89.

Yahnke, C.J., Gamarra de Fox, I. and **Colman, F.** 1998: Mammalian species richness in Paraguay: the effectiveness of national parks in preserving biodiversity. *Biological Conservation* **84**: 263–8.

Yen, D.E. 1989: The domestication of environment. In Harris, D.R. and Hillman, G.C. (eds), *Foraging and farming:*
the evolution of plant exploitation. London, Unwin Hyman: 55–75.

Young, E. 1992: Hunter-gatherer concepts of land and its ownership in remote Australia and North America. In Anderson, K. and Gale, F. (eds), *Inventing places: studies in cultural geography.* Melbourne, Longman Cheshire: 255–72.

Young, K.R. 1998: Deforestation in landscapes with humid forests in the central Andes: patterns and processes. *Nature's geography: new lessons for conservation in developing countries.* Madison, University of Wisconsin Press: 75–99.

Zimmerer, K.S. 1994: Human geography and the 'new Ecology': the prospect and promise of integration. *Annals of the Association of American Geographers* **84**: 108–25.

Zimmerer, K.S. and **Young, K.R.** 1998: Introduction: the geographical nature of landscape change. In Zimmerer, K.S. and Young, K.R. (eds), *Nature's geography: new lessons for conservation in developing countries.* Madison, University of Wisconsin Press: 3–34.

Zimmerer, K.S. and **Young, K.R.** (eds) 1998: *Nature's geography: new lessons for conservation in developing countries.* Madison, University of Wisconsin Press.

Zvelebil, M. 1994: Plant use in the Mesolithic and its role in the transition to farming. *Proceedings of the Prehistoric Society* **60**: 35–74.

Index

Note: page numbers in **bold** refer to figures and illustrations; page numbers in *italics* refer to tables and boxes.